HURRICANE COMBAT

HURRICANE COMBAT

The Nine Lives of a
Fighter Pilot

Wing Commander
K.W. Mackenzie, DFC, AFC

Foreword by

Air Chief Marshal Sir Denis Smallwood
GBE, KCB, DSO, DFC, FRSA, FRAeS

WILLIAM KIMBER · LONDON

First published in Great Britain in 1987 by
WILLIAM KIMBER & CO. LIMITED
100 Jermyn Street, London SW1Y 6EE

ISBN 0-7183-0633-3

Photoset in North Wales by
Derek Doyle & Associates Mold, Clwyd
Printed & bound in Great Britain by
Redwood Burn Limited, Trowbridge, Wiltshire.

Contents

List of Illustrations

The author and publisher would like to thank Mr Chaz Bowyer for his help in providing some of the photographs listed above

ILLUSTRATIONS IN THE TEXT

Foreword

by

Air Chief Marshal Sir Denis Smallwood
GBE KCB DSO DFC FRSA FRAeS

The first prototype of the Hurricane first flew from the airfield at Weybridge on 5th November 1935. Exactly half a century later, I attended the 50th Anniversary celebrations hosted by British Aerospace at their Weybridge factory. Many distinguished former operational Hurricane pilots were present, including Air Chief Marshal Sir Harry Broadhurst and Roly Beamont of TSR2 test pilot fame. During World War 2 most, including myself, had spent a good deal of their operational time flying Hurricanes in the early part of the War, and then Spitfires later. All spoke with highest praise of the Hurricane's operational characteristics, particularly as a stable gun platform. Apart from being the major instrument of victory in the Battle of Britain, there was no other single seat fighter which served so successfully in such a variety of operational roles throughout the war, including night fighting, night intruders and tank busting. All felt, however, that whereas the Spitfire had received full acclaim and even adulation, much more could and should be said and written about the Hurricane. Ken Mackenzie's book, apart from being very readable and exciting, goes a long way towards achieving this aim and gives a first hand detailed account of his time as an outstanding operational pilot with the Hurricane in 1940/41.

For me, having been closely connected with the Hurricane in that same period, there is much upon which I would like to comment and elaborate. I will confine myself, however, to making a few observations on the ill-fated raid on Lannion, described by Mac so vividly in his first and eleventh chapters, and make a few comments on the planning and tactical aspects.

This ill-fated mission, as the reader will soon discover, is the turning point in Mac's book, as his brilliant operational career was abruptly terminated – to be faced then with the gruelling and despairing life of a prisoner of war. The raid on Lannion was, to my mind, a classic example of an ill-conceived, badly planned operation, based on faulty intelligence. Apart from the fact that we had no recent practice in this sort of offensive operation, such was the short notice given to us by Sector Headquarters that there was no time to discuss tactics in any detail, let alone to consider any maps or diagrams of the precise airfield layout at Lannion itself. 'Y' service intelligence had reported a large reinforcement of Heinkel 111s to Lannion with the intention presumably of carrying out a heavy raid on one of the major cities in the West of the UK. To employ such a small force of Whirlwinds, followed by a mere four Hurricanes to counter this formidable threat was pitiful indeed; to send in the Hurricanes well after the Whirlwinds, thereby losing all element of surprise, was sheer stupidity. When the four Hurricanes eventually arrived over Lannion, just before last light, the German defences were fully at the ready and the reaction was immense. The fact that Mac was the only casualty, and that three out of the four managed to limp home riddled with shot, was nothing short of a miracle. As Mac emphasises, there was no sign of the Heinkels, reinforcements or otherwise. Apart from demonstrating to the Germans that the Royal Air Force was maintaining and increasing its offensive on all sectors, the raid achieved little or nothing of value to the war effort.

It is also interesting to consider the tactics we adopted for this operation and to note that not much has changed in modern day terms. The tactics used in the Falklands campaign for ground attack were basicaly the same – i.e. go in as low as possible, and as fast as possible, and then get to hell out of it, and above all do not be tempted to make a second pass (Mac freely admits that it was his second pass, prompted by his pugnacious and press on spirit, which led to his downfall).

In the Falklands, the Harrier was the major instrument for ground attack. In one of the first ground attack missions, the pilots were briefed to penetrate the target area at some 520 knots which was the speed calculated in flying trials to produce the best weapon delivery accuracy. In debriefing after the mission, one pilot reported that he had penetrated at a maximum speed of well over 600 knots. On being asked why, he replied, 'Because it wouldn't bloody well go any faster!' Tout ça change, plus c'est la même chose!

I will not attempt to comment on the latter half of Mac's book and the terrible time that he had as a prisoner of war. I would only say that when I met him again some forty-five years after the Lannion raid, I might well have expected to find him looking rather ancient and to have slowed down considerably, after all he had been through. But no, not a bit of it; he looked much the same, and that fierce spirit of determination, that press on vigour were there just as forcibly as I had grown to know and admire them in 1941. Old airmen never die. They don't even seem to fade away.

Dedicated to the personnel of No 501 (County of Gloucester) and 247 (China British) Squadrons and all those pilots in the Battle of Britain who were less fortunate than I was.

CHAPTER ONE

The Intruders

September 29th 1941 was one of those days. Briefed to attack Lannion airfield in Brittany, we took off at dusk from Predannack, the most south-westerly aerodrome in England at 18.50 hours to follow a first strike by Whirlwind fighters of No 263 Squadron. It was with mixed feelings that we lifted off in four Hurricane Mk IIc's as we were quite convinced that there were few, if any, good targets at Lannion at that time.

As we dipped down onto the sea from Predannack to cross the Channel at about 200 feet in order to get under the German radar, we spread out into two flights of two aircraft on a course of 152 degrees, throttling back to a cruising speed of 230 knots with Red 1 and myself (Blue 1) slightly ahead of our No 2's.

The weather was hazy with some high stratus cloud and the sea looked dark and, as usual, cold and uninviting, but calm. Keeping radio silence we skimmed over the water, settling ourselves into our seats with tight harness and well positioned, relative to the gun sights: the Merlin engines were droning steadily at high boost and low revs for best cruising speed and economy before re-adjusting for the final attack. Looking across at the other aircraft we looked like black crows with the all-black matt finish and only a slight glow from the stub exhausts in the fading light and haze. After some 25 minutes we made out the French coast with lower cloud inland drifting towards the target area. Passing Les Triège Islands to our east we selected full power, checked all instruments, switched on the Gyro gun-sights and guns to 'Fire' and pulled up to 600 feet for the diving attack on Lannion.

Flying in line abreast and 200 yards apart, we dived to attack at

100 feet; Red 1* and 2 attacked the western perimeter and I, with my No 2†, took the east side. As we dived to attack the flak defences opened up immediately with intense and accurate fire from all sides of the airfield and outside it. Visibility was poor, even with our canopies open, in the hazy fading light so that the tracer bullets of the flak showed up like a stream of fireworks, only more deadly. Red 1 and 2 fired long bursts into what looked like dispersal pens without any visible result.

With my No 2 we weaved and fired at buildings and dispersals; I silenced a gun post about half a mile from the north of the aerodrome. The flak was a constant stream by now and getting more accurate, missing me by a few feet only as I jinked and could only get in short bursts of fire at the guns. Red 1 and 2 and Blue 2 had pulled away, but I was determined to have a last quick attack on my way out towards the coast.

By now the sky was alive with flak and I was sweating like a pig and very angry about the whole mission. I silenced one more gun position but was coned by the rest as I passed to the south of the airfield down to about fifty feet to upset their aiming, but too late and I felt the thud of bullets hitting the aircraft, all – thank God – in front of me, in the engine, radiator and oil with one across the top of the windscreen and a crack on the top of my head. I pulled straight up, nursing the engine. Height, I must get height as a means of escaping towards the sea and base. Visibility was now very poor with long dark shadows and haze. With a failing engine and a sore head, I continued weaving as the flak fell away out of range though still following me in vain.

My aircraft was obviously in a bad way: engine rough, oil fumes, banging and clattering, power falling off, as I eased back on boost and revs as the temperature gauges rapidly rose to dangerous levels. I was going to have to fight for my life if I was to get back to base, or even out of this stricken aircraft. Clearly I was not going to make it as, at around 1,800 feet, the engine slowed desperately and the smell of hot metal warned me against fire, the fighter pilot's greatest danger with a large gravity tank in front over one's feet.

Ditching or baling out was going to be necessary, but as far from

* Red 1: Flight Lieutenant D. Smallwood; Red 2: Pilot Officer Hordern.
† My Number 2: Sergeant Deuntzer.

the coast as possible and which and where? I levelled off, cut the engine back to a tick-over – more like a 'banging over' in fact – and trimmed the aircraft into a gentle glide at 120 knots on a course of 320 passing over the coast which was only a dim outline with a few scattered lights here and there. Ahead, the sea was barely visible as such. I quickly thought about baling out versus ditching. I had already baled out once and did not like the prospect of doing so again – especially into the sea – although I had already lost some front teeth in a forced landing in the Battle of Britain the prospect of ditching seemed the best bet.

All this flashed through my mind when the engine decided to seize with a grinding clatter and the prop stopped and I had to retrim to keep my gliding speed up. Suddenly the thought of ditching, with a large protruding radiator underneath to drag one under or possibly pitch the aircraft over onto its back, crossed my mind ... I was in fact quite unprepared for the ditching qualities of the Hurricane.

With the sea rising rapidly as I glided down I hastily jettisoned the hood, tightened my harness, peering ahead for the sea, now about five miles from the coast. Checking the oxygen full on I selected full flap and switched the landing lights 'on' to better judge my height as we settled towards the sea. About six feet up I stalled the aircraft onto the water at about seventy knots. She struck immediately, slowed rapidly and pitched forward violently, water and spray everywhere, with loud banging and hissing from the rapidly cooling over-heated engine.

It was an uncanny silence that followed as I undid my harness, broken by the hissing and crackling and steam from the engine, took off my helmet and eased myself up out of the cockpit as she seemed to be nosing under quickly. This was my last recollection of my actions until I came to some time afterwards floating round in my 'K' type dinghy, wet through, sitting in about three inches of water, cold as ice, dazed but alive and all in one piece with a sore and bloody head. No sign of my helmet, parachute in its pack or the aircraft, very much alone in the dark calm sea, dim reflections and lapping of the water on the sides of the dinghy ... a very small dinghy indeed, barely long enough to stretch out in, with totally inadequate paddles. In short, 'up the creek' but at least with minute paddles and alive.

CHAPTER TWO

Beginnings

So, what now? Where had it all started?

Born in Belfast, I was a keen sportsman, but no great scholar unless really interested in a practical subject. Playing a lot of rugby and enjoying some success in athletics, I suppose it was natural that flying and cars attracted me from my school days. Making and flying model aircraft became a craze and modifying and tuning the family cars an absorbing pastime until 1934 when Alan Cobham, as he was then, visited the Sydenham area near Belfast with his famous Flying Circus.

This was by way of being quite an historic event for Belfast, as the area used was destined to become Belfast's first civilian airport at the head of Belfast Lough and adjacent to Short and Harlands. The visit and display of flying gave aviation in Northern Ireland a boost that was instrumental in furthering flying in Ulster and encouraging the formation of the local Volunteer Reserve in good time to augment the Royal Air Forces position at Aldergrove with 502 Auxiliary Squadron.

Together with many others I invested 7s 6d – or was it 5s 0d – in a short flight in one of the Circus Avro 504's. From that moment I was convinced that to fly, zoom and soar like a bird, be part of the elements, the wind and clouds, dive and climb and control such a machine was what I most wanted to do, the exhilaration, individualism and freedom were indisputable.

Starting an engineering apprenticeship at Harland and Wolff before going to Queens University to study for an Engineering Degree, I began to wonder if a career combining both could be envisaged. There was little thought of war at this time, though Germany's rearming etc., was well publicised.

Therefore I was overjoyed to hear that Airwork Limited were opening what was to become the North of Ireland Aero Club at Newtownards airport at the head of Strangford Lough, the western boundary on the famous 'Comber' straight – part of the Ulster TT race course. With support from my parents I became a founder member of the club and commenced flying as soon as possible.

The Chief Instructor was Mr Bryant, ex-RAF, who brought the RAF standards of instruction into the civilian club programme and sequences. We were also lucky in enjoying the patronage of Lord Londonderry, the then Under Secretary for Air, whose estates covered most of the local area. Some of his family also became members and learnt to fly there. The club was equipped with Avro Cadets with 215 HP Siddeley Lynx engines, similar in general specification and performance to the Avro Tutors used in the RAF for Elementary Instruction, except that the Cadets did not have the inverted flying modifications of the Tutors. Larger than the De Havilland Moth, the Cadet was more powerful with a 'bigger' aircraft feel and better performance in every respect, though possibly more forgiving than the Moth in some respects.

With only a few members, and being only three-quarters of an hour away from Belfast by road ... half an hour if you belted it! ... and several aircraft available, there was plenty of scope for rapid progress and continuity in training, the latter essential to save wasting time going back over one's last exercises.

I went solo in six and a half hours and thereafter enjoyed halcyon days and weekends of intensive training and instruction in all the facets of elementary flying and navigation, culminating in getting my 'A' licence and graduating to cross-country flying, aerobatics from an early stage into planned display sequences. I spent much time in the hangar with the servicing mechanics learning the maintenance schedules from them, delving into construction, watching, listening and helping when allowed.

One of the highlights of club life was the decision in 1937 to include Newtownards in the King's Cup Air Race route. Acting as hosts we had the opportunity to see and meet many of the famous pilots of the time, inspect their various aircraft, such as the De Havilland TK4, flown by Waight, Alex Henshaw's Miles Hawk and Percival's Mew Gull, the Miles Falcon, Comper Swift, etc., to name but a few of the many competitors and aircraft. The almost total lack of forward visibility of several made us wonder how they landed them. To find out, we blanked off the screen of the front seat in a Cadet with cardboard and practised landing looking over the side ... it worked!

With a natural bent for things mechanical and sporting, I became completely involved; navigation exercises, cross country flights, competitions with spot landings, bursting balloons, etc. The CFI encouraged me, rather against his will I believe, to perfect aerobatics and turned a blind eye to blatant showing-off over Bangor Bay at weekends, albeit within the regulations regarding height and distance, etc. This resulted in sympathetic support from the CFI to begin working towards an Assistant Instructor's Ticket. I was, of course, elated at the prospect.

I was by now doing Engineering at Queens University and was a member of the OTC (Royal Corps of Signals). The Civil Air Guard training was under way in England and the Volunteer Reserve Schools well into their stride, whilst the Auxiliary Squadrons were expanding and being re-equipped.

War clouds were gathering. Some of my friends suggested that I should join 502 Squadron (Auxiliary Air Force) stationed at RAF Station Aldergrove. This seemed a good idea at the time, though the prospect of flying heavy lumbering Heyford bombers did not appeal to me: I felt that my character and experience to date were more suited to fighters. Anyway, an application form was made out with a brief CV (not much to go on it at this stage), forwarded and an interview arranged.

Prior to this, I had bought my first car, a 1932 Wolseley Hornet Swallow two-seater, which we had converted from a pointed tail to an 'MG' type slab tank tail with luggage space. Total cost was about £20 to buy and modify! After a year's enjoyment this was exchanged for a 1934 MG 'PA' owned by a friend of mine, which we had assisted in tuning, preparing and running when he came third in Phoenix Park, Dublin, before joining the RAF. In this I presented myself at Station Headquarters, Aldergrove, and was ushered into the Adjutant's office and thence to the Squadron Leader Administration's office.

Having had two years' University OTC training as an officer

cadet, I had a sound grounding in discipline, but also had a fair share of the Ulsterman's distrust and intolerance of 'bull', with a total abhorrence of pomposity and snobbishness. Therefore, my first rub with the regular Air Force in the shape of this tubby, unfit, little squadron leader interviewing officer came as what we call a 'stunner'. His manner, facetious questions, lack of local knowledge and contempt for civilian flying and general attitude put me off immediately.

After preliminary introductions, this over-weight officer asked why I wanted to join the Auxiliary Air Force. I told him. He asked why had I learnt to fly at the North of Ireland Aero Club. I told him! He asked what flying experience I had, despite having all the relevant details on the Application Form in his right hand. I told him. 'Ah!' said he, 'that's only civilian flying, not RAF!' I pointed out that the CFI was ex-RAF. He was not impressed ... neither was I. He asked did I have a bank account; was that my car outside; what did my parents do; why was I at Queens University? I thanked him for his trouble, took my application form from him, thanked him for wasting my time and left him spluttering in rage. A happy escape as it transpired, as in a short time I learnt that 502 had been absorbed into Coastal Command.

As the Volunteer Reserve had opened a Town Centre in Belfast and a flying school at Sydenham – now fully established as an airfield in the form of No 4 Elementary and Reserve Flying School run by Short and Harlands, whose shipyards were adjacent and Administrative buildings on the road to the airfield – I decided to try there. The war was now quite obviously 'on' and it was only a matter of when and although there was no conscription in Northern Ireland, our loyalty to the Crown was traditional and the response to join the various local units of the three Services was already extensive.

I joined the Volunteer Reserve and was welcomed with open arms, having some basic flying experience. The Chief Instructor was Flight Lieutenant D. Sloan (seconded from the RAF), assisted by Flight Sergeant W. Lake, both RAF qualified flying instructors and ex-CFS (Central Flying School), with Flight Lieutenant Lindsay from Aldergrove as required. We soon had over forty under training, starting on De Havilland Moths and graduating to

Hawker Variants, Hind(dual), Audax and Harts in a splendid unit. Being close to Belfast the location was ideal for reasonably intensive continuation training.

Much has been written about the Volunteer Reserve: the fact that all entrants were initially sergeants; that resentments were rampant when they were absorbed into regular units; that jealousies existed on squadrons later; that there was discrimination and other such nonsense. In fact it was what you made of it. Before the war there were probably three distinct types of aircrew. The Regulars made up of career men, professional air force men; others who joined because they wanted to fly and many who joined because they were not qualified and frankly, jobs were scarce. In our opinion these were all tied to a system; somewhat hidebound procedures and traditions, dependent upon the RAF for their livelihood, career conscious at all times and therefore restricted in thought and deed in many respects.

The Auxiliary squadrons, with full squadron status and aircraft establishment administration and station facilities and personnel were rather unique. The pilots were mostly men slightly older than their counterparts in the regular service or the Volunteer Reserve who were established in business or could afford the life anyway, but again all men who loved flying and the life. Then came the Volunteer Reserve, conceived for rapid expansion of the pilot and navigator establishment necessary for the emergency. They were a very mixed bunch of individuals, from all walks of life and social background, with one common factor – a keenness to fly, either because of their love of flying, or because of the threat of war and a desire to get into it early. Most were either professionally qualified, training to be, or in jobs already and not dependent upon the RAF (VR) for a salary, albeit the weekly attendance money was welcomed, of course. The Auxiliary and VR units were therefore uninhibited, not career-conscious or restricted by the system of procedures, free to express themselsves in a common bond: an interest in flying.

For myself, it was what I had planned to do – to make flying my career, combined with engineering if possible, either civilian or RAF – but first there was a war coming up and we looked forward to it rather as we did the next rugger match, without any thought to the political or national implications. Indeed, politics was not an issue as it has become now and at that time Ulster was a stable place, where freedom was cherished with rabid individualism.

At Sydenham, under David Sloan and his team of flying instructors and Town Centre staff, together with the Short & Harland servicing crews, we soon had as near a squadron atmosphere and spirit as possible, without the hassle. Morale was high, attendance intensive and the instruction could not be faulted.

David Sloan was almost the 'typical' RAF type – extrovert, with a ginger moustache, ruddy complexion and infectious manner. He and his wife joined into the VR spirit with gusto and whilst preserving an apparently easy-going attitude, retained a nicely balanced official grip on us all. A patient and efficient instructor, he inspired confidence in his pupils, but suffered fools with a quiet discipline. Supported by Bill Lake, a likable, sardonic, vastly experienced flight sergeant of the best possible type, they were a marvellous combination to handle our mixed bag of Ulstermen ... we felt that they were 'one of us' and that meant a lot in those days. It is to their credit that the VR in Belfast never lost an aircraft or a pilot under training. The nearest accident was a Hillman Airways De Havilland Rapide which went on fire in the air in the circuit. The pilot jumped out without a parachute and both crashed on our perimeter in soft ground that bogged down our fire crash tender. I suppose that fire was the one danger we feared the most ... or did not think about.

Our Town Centre instructors were either seconded RAF or retired RAF, also a grand team. We were lucky indeed. The servicing crews were mainly local Short & Harland engineers with a couple of ex-RAF retired senior NCO's who had years of experience on the Hawker variants and Rolls Royce engines. They treated us with amused tolerance: ' ... sergeants indeed??? ... took us b ... years to get our stripes'! ... but it was said without rancour and they welcomed both our visits to the hangar and interest in the servicing. We learnt a lot in those days which brought us closer to the aircraft and the men upon whom our lives would depend. My respect for the ground crews was born then and has never wavered, their efficiency and dedication enhanced by a sense of humour found nowhere else.

Being so close to Belfast and the airfield most of us enjoyed and undertook frequent attendance, flying as often as possible – even at lunch time – attending ground lectures and examinations interspersed with social events, rugby matches, dances, car rallies. Flying discipline was excellent and the number of incidents minimal; in fact the accident record was exceptional indeed. My previous flying on the Avro Cadets stood me in good stead and after a minimum of refresher flying to the RAF syllabus and 'patter' on the Moth I graduated onto the Hawker variants.

Once on the Hind Trainer, which differed from the Hart and Audax by way of its fully supercharged Rolls Royce Kestrel MK V engine with stubb exhausts, flying took on a new dimension. Here was power, an aircraft to be used ... an extension of oneself, to be at one with ... like a thoroughbred horse which responds to usage; an aircraft to be flown to the limit. Not a forgiving aeroplane ... it could bite back ... and did ... if abused ... or one to take liberties with, but one which when flown well was exciting, satisfying and could be exploited to the full. Aerobatics, spot landings, dog fighting, height climbs, navigation exercises, forced landing practice, side-slipping off height, night flying ... you name it; with all the exuberance and confidence hitherto unknown. I revelled in it and spent more time than I should have at Sydenham, to the detriment of my studies: flying, with the servicing boys in the hangar, helping with the administration.

I resigned from the North of Ireland Aero Club and the OTC and gave up any ideas of becoming an assistant instructor at Newtownards to concentrate on the VR; being made sort of 'head boy' and encouraged again to fly as often as possible. In this way I was lucky and able to get in many hours' concentrated training including cross country flights to Scotland and England, to Prestwick, Sealand and South Cerney and met many regular trainees on Hawker variants. It was still a carefree existence in many ways, although the threat of war was looming ever closer. The VR pay kept us in pocket money and paid all expenses. The little MG 'PA' was rallied and used to and from the aerodrome where somehow it managed to have a full tank after each weekend; mildly tuned with larger SU carburettors it could do a genuine 75 mph.

We received several more Harts and Audaxes ex-North-West

Frontier in early 1939 and as these were brought onto the line we air-tested them. Whilst testing one I found the aileron controls very sloppy and landed to have them looked at. When finally parked in the hangar and the engineers opened the inspection panels in the fabric on the wings we were swamped in a cascade of table tennis balls! The bottom wings were also full of them and so were the wings of all the ex-Indian aircraft; apparently they had been put in for buoyancy should they have to ditch on the flight back to UK ... most were sold to local sports stores.

By this time training was becoming more earnest and I was flying some sorties as target for the local anti-aircraft units ... no live ammunition, of course, this was quiet fun flying requiring accurate height and course-keeping.

In July 1939, I was told by David Sloan that the airfield was going to be officially opened as Belfast Airport on 31st July by the Under Secretary of State for Air, Captain H.H. Balfour, that all the VR together with their aircraft were to be on parade, that I was to give an Aerobatic Display and be presented by Captain Balfour with the Short and Harland Cup for the 'best all-round trainee'. Naturally I was quite taken aback but at the same time pleased and honoured. There was to be a full rehearsal some days beforehand weather permitting, and I was to work up a suitable aerobatic sequence and practice ready for the rehearsal. Nothing like being ordered to do aerobatics over base at a satisfactory height to do the aircraft justice!

In the event the opening went very well, the weather was kind and the display a great success, the Hind being ideal for a 'tight' display well within the perimeter of the airfield and when the Sunderlands failed to arrive due to fog I was asked to do a repeat, followed by flying one of the VIP's over to Scotland later. The cup is a treasured memento to very happy days.

Some weeks later I was again coming back from Prestwick one evening and found Belfast Lough shrouded in low sea fog mingled at the top end, where the Sydenham airfield is, with the local smog from Belfast's factories etc. The fog was only about 300 to 400 feet thick with the tops of some of the shipyard cranes protruding through and chimneys of the power station visible at times, but Sydenham, on reclaimed land on 20 feet or so above sea level, was obscured with no land marks or pinpoints for guidance. The two diversion airfields were either Aldergrove or Newtownards and I did not fancy either of them, so decided to go back to the entrance to the Lough, let down to about 50 feet and fly up using the shipping lane buoys as guidance. I flew back and let down beyond the fog and where it was clear and turned up the Lough at 50 feet with a forward visibility of a quarter to half a mile, knowing that I only had to zoom up 500, or so, feet to be in the clear again. Keeping station on the buoys, I completely forgot about the pilot boat at its moorings.

I should perhaps clarify that as Sydenham was reclaimed land it was only about twenty feet above sea level and one could 'lift up' onto it, as it were, off the sea. The pilot boat was quite large and moored in the Lough to supply pilots to incoming boats and collect them off outgoing ones, and was a sort of permanent fixture. Anyway, with airspeed down to about 100 or so I crept up the Channel, suddenly saw the pilot boat ahead, banked violently and narrowly missed it and continued to find Sydenham 'on the left at the top end' there literally dropping straight in to land, making a 'wheeler' as my speed was too high to round out and make a three pointer, and in fact stopping at the boundary fence. The pilot boat reported some 'BF' in a silver biplane nearly taking their masts off. I was the 'BF' and should have diverted to Aldergrove or Newtownards; a salutary lesson in over-confidence.

In early August we were mobilised. The dogs of war had been released on the Continent, but war had not yet been declared. We reported daily to the town headquarters, flying was virtually stopped and the VR staff were taken by surprise. We coffee-shopped, helped with administrative work and then had word that all the aircraft were to be flown to England and dispersed around Service flying training units with similar types. Meantime the obvious thing it seemed to us would be for the Air Ministry to take those with a certain level of flying experience and number of hours for immediate final training – many were at Wings standard already – and then concentrate on the very many others and hundreds of volunteers flocking to join, but without any flying experience. How wrong we were. This was too logical and sensible, as we subsequently found out to our chagrin and frustration.

For the time being we attended lectures and I stepped up my attendance at Queens, but was convinced that it was 'any day' now. We tested aircraft as they were serviced and helped at Sydenham; but no flying. September 3rd came and the declaration of war – this was it! … or so we thought, and we arranged a full VR farewell luncheon at the Grand Central Hotel, with appropriate speeches, barrels of Liffey Water and a sprinkling of Old Bushmills. Nothing came of it; no postings as expected; we languished. The monotony was finally relieved by ferrying the many aircraft over to England, collecting our pilots in our lately acquired Anson 1, which having some twin time on a Monospar and Q6 from the Aero Club, I was able to fly. This also made me more certain that I had no desire to go on to bombers or coastal aircraft.

Impatient to get into the fray we wasted days, into weeks, into months, being paid fortnightly and getting more bloody-minded daily. In November we bearded our VR Commanding Officer in the Town Centre with a terse request to 'do something'; whether he did or did not, our postings came in in late November. These were for us to report to various Initial Training Wings in Hastings … or nearby. We duly reported, but were in no mind to be b – about … Of course we were, in no uncertain fashion.

CHAPTER THREE

Frustration

The Initial Training Wings on the south coast had been formed to take Royal Air Force Volunteer aircrew under training in preparation for posting to the various Service Flying Training Schools (STFS's). They were an expedient, hastily opened without much planning or thought; that they succeeded at all was due mainly to the staff involved and the spirit of the trainee, but owed nothing to foresight by the Air Ministry, or imagination by the senior staff administration.

I arrived at Number 3 ITW Marine Court, Hastings on December 28th, together with several hundred others. The place was a shambles, bare, dirty, badly lit. We scrambled for the smallest rooms with straw-filled mattresses on iron bedsteads, bare rooms, lifts out of order, smelly and cold. We marched everywhere, queued for the toilets, bathrooms and food – the latter laced with bromide! We were in neither the mood nor the state for female company anyway! Every available public place on the promenade, in the town, seemed to be used for marching, drilling, PT; the cold was intense; a very heavy winter.

We attended lectures, played rugby, squash at the local club, but did absolutely nothing constructive related to flying; a shocking waste of the previous year or so to many of us. It was a social leveller; we were mixed with those who had never flown and, rightly or wrongly, we resented it like hell.

In short, it brought us down to earth with a vengeance and we had to 'thole' it. For myself it did not last too long, as after a few days I was made a squad commander of about sixty. This had some privileges and status amongst the regular senior NCO's and staff and relieved the monotony as it entailed disciplinary and administrative work. At the same time, and through this work, I got to know many of the others with considerable previous flying experience, including several ex-instructors, 'B' licence holders and a couple of ex-airline pilots. We discussed our plight and formed a small clique from various billets along the sea front and became determined to make some effort to end this, to us, useless 'Hastings' holiday.

Meantime, over in France the Advanced Air Striking Force of the RAF was getting some action. We decided to present a petition, signed by all concerned, to go to General Critchley, who commanded these training wings ... to us diabolical units, where we now had an outbreak of 'flu to beat all 'flu. I suppose we were a rebellious lot, not the least being myself. The petition was duly presented and several of us were called upon to answer for it. This we did in forthright, but not disrespectful terms, as we appreciated that by then the severe winter was playing havoc with flight training and the SFTS's were probably well behind schedule in consequence. Nevertheless, this was also an argument to make the most of our experience and hours to fill the gap. In short, our plea had been overtaken by events, but they promised to look into it. At least we had made our point.

To enlighten our existence and promote some interest in the main objective, namely flying, Ann Welch and some of the others of the gliding fraternity had been contacted and they arrived with several gliders and towing winches to set up a small gliding centre on the downs south of Hastings. Several of us were involved in this and excused the routine drill, PT and lectures; at last a step in the right direction.

In March our prayer, or petition, was answered and most of our clique were posted, not as we had hoped to SFTS's for advanced training, but to Elementary Flying Schools! We made one final protest, to no avail of course. I went to Hanworth on the Great West Road outside London, with the old Manor House almost in the middle of the airfield which had first been associated with aviation in the First World War. Its fame was enhanced by the arrival of the Graf Zeppelin in 1931, followed by the location of various aircraft designers and manufacturers there, together with several training schools. Some succeeded and some failed, culminating in the establishment of Number 5 Elementary and Reserve Flying Training School there in 1935, with Blackburn B2's very similar in specification to the De Havilland Moth. At the outbreak of war, the unit was re-equipped with Miles Magisters, whilst General Aircraft Ltd. experimented with their ill-fated Fleet Spotter, with four small Pobjoy radial engines, which we were to see from time to time.

At Hanworth we were introduced to the Miles Magister, an all wooden aircraft, a monoplane with more modern concept and flying characteristics, including flaps. It did not endear itself as the Cadet and Moth had done and the general handling and performance, though updated, were mediocre and did not inspire confidence.

Our life was made easier by the Chief Ground Instructor, Captain Oldmeadows, a wonderful character with one leg, the other lost in a flying accident, to whom nothing was too much trouble and every day a cheery one. He sympathised also with our plight, especially with the ex-civilian commercial pilots. At least we were back on flying and another slight protest was submitted with the result that our stay was a short one I was passed out as 'above average'. The highlight came on the last day when a student skidded into a line of parked Magisters, writing off about seven of them.

We had managed several visits to the dimmed lights of London, seen a show or two and enjoyed a little freedom. The flying was frankly boring and repetitious, and lacking in inspiration for pilots destined for operational squadrons, due to the fact that the instructors were mainly ex-civilians, mobilised into the VR themselves. Travelling by public transport during this 'phoney war' was slow and difficult, so I invested in a 250cc 'OK Supreme' motor cycle on sale in a nearby garage and resolved to bring the MG 'PA' over from Ireland as soon as possible.

Posted from Hanworth to Number 3 SFTS at South Cerney, I took the 'OK' down with me and was glad to meet some of the staff still there whom I had met previously on cross-country flights from Sydenham. This was more like it! Arriving in May, in a period of exceptionally good weather, we hoped for a quick refresher and then to a squadron ... a forlorn hope. However, at least our previous experience and assessments were acknowledged and after very brief check flights we were shown some latitude, enjoyed good flying, took our Wing examinations and several of us were moved into the officers' mess as officer cadets ... getting somewhere at last.

The flying built up to quite an intense rate including cine-gun and night flying from the satellite aerodrome at Bibury, just above the famous Swan Inn, which still produced magnificent steaks in 'olde worlde' surroundings in a picturesque village. The war seemed far off, deep in the West Country as we were, with Cirencester on our doorstep. The atmosphere was conducive to high spirits and the flying produced good results. The German Blitz through the Low Countries and France was alarming and enemy air activity increasing so that we were more than eager to 'get stuck in'. I was asssessed 'above average' again on Hart variants and recommended for fighters whch made me cock-a-hoop, not caring which, Spitfires or Hurricanes.

Before leaving South Cerney, however, I was returning to the Mess on the 'OK', when turning into the officers' mess drive, I forgot the large roundabout in front of the main entrance ... and rode straight through the rose bed leaving a trail of broken rose trees behind me and coming off at the far side sore and covered in thorns. The next day was spent replanting and digging the bed over, losing a day of my leave ... fair retribution!

CHAPTER FOUR

Fighter Conversion

Before reporting to Number 6 Operational Training Unit at Sutton Bridge, on the Wash, for conversion and gunnery training, I sold the 'OK' – as I had had an offer for it – and took the remaining four days' leave to go over to Ireland. I collected the MG 'PA' off its blocks and motored to Belfast and onto the mail boat in good time – with the help of Mr Grogan, the Chief Purser, and a full tank of petrol from some of my garage-owning friends – a successful sortie.

A safe arrival in Liverpool was followed by a glorious fast drive on clear roads in good weather to Sutton Bridge, scrounging a few gallons of 'pool' petrol here and there on the way. Now resplendent in pilot officer's uniform and arriving a day early, I was quickly introduced in the mess and made welcome. There were several instructors who had already been 'blooded' serving with the AASF in France, including Billy Drake, DFC.

The next day I was introduced to the Hurricane. This was brief and to the point, with a short cockpit check on the operation of the retractable undercarriage and flaps in one of the aircraft on trestles in a hangar. It was in fact an updated Hind, Hart or Fury without the top wing, with a closed cockpit and retractable undercarriage and more powerful Rolls engine. A fully grown up monoplane Hawker, just asking to be used.

Before flying the Hurricane we were given a quick check flight in a Harvard I. My instructor was a Flying Officer Jackman; this was to accustom us to the lack of forward vision on the approach and round out and to demonstrate a 'fighter' circuit and the use and effect of the flaps and retractable undercarriage.

After two short flights I was briefed for my first Hurricane solo. It was glorious. Here was an aircraft that could be danced in the air, flown to the limit, tight-circuited and side-slipped like a Hind or Hart, inspiring great confidence and encouraging the pilot to become as one with it. Responsive, strong, stable and forgiving, except for spinning, which required considerable height in hand for recovery.

We flew intensively; formation, height climbs, aerobatics, experimenting with the tightest possible turns ... 'greying out', recovering, sustaining the turn until 'blacking out'; learning to tighten the stomach muscles, tensing up, to increase one's 'black out' threshold. We practised pairs dog fights, formation aerobatics, air firing, sighting, cine-gun pairs flying, getting the feel of the guns, spot landings, night flying and instrument flying. Flying the aircraft to the limit; the only way to become master of it, as one with it. It was exhilarating and the average of three sorties a day never seemed enough.

Using the guns showed us what a stable gun platform the Hurricane was, but those of us with a firearms' background immediately wondered about the adequacy and effectiveness of the 8×303 calibre machine guns, small bore short range weapons with no real weight of metal, no hitting power. We felt that spot harmonisation at, say, 250 yards, would be more effective than the standard setting. However, being new at the game, we were a voice crying in the wilderness; we had to accept it, but our misgivings were proved to be right later in squadrons in the Battle.

The general lack of experience in shooting, using the guns, using the aircraft as a weapon and the inadequacy of the 303, except at close range, was to be the greatest shortcoming in Fighter Command during the Battle of Britain. An additional factor was that 'limit flying' had not been introduced or taught before the war: that is, getting the maximum performance out of an aircraft compatible with the configuration and loading at the time. In short, using it to the absolute maximum of its capability, taking it to the limit of its construction and flying capabilities and those of the pilot himself. Thus many pilots, including instructors, did not get the best results from their aircraft or weapons. This aspect of flying has, of course, been introduced to good effect worldwide since the war.

Mess life at Sutton Bridge was lively as, between flying such a magnificent new aeroplane, men were comparing experiences, performance, what to do, when and how. A new life was born there for most of us; a few fell by the wayside. One unfortunate stupidly stalled on his round out over the dyke onto the airfield, ended up on his nose, climbed out of the cockpit uninjured and, slipping on oil spilled from the tank in the port wing, fell and broke his leg!

We now looked forward eagerly to news of our postings, hopefully into the Battle area. For myself alas this stage was yet to come – not quite, anyway.

CHAPTER FIVE

Nearly There

On September 20th, with the Battle of Britain into its third phase, I was posted to No 43 Squadron, the famous 'Fighting Cocks', lately withdrawn after a mauling at Tangmere, to Usworth. There I teamed up with George Westlake, under Flight Lieutenant Frank Carey, DFC and Bar, DFM, a very experienced pre-war fighter pilot. Squadron Leader T.F.D. Morgan, DFC, was our commanding officer, who had taken over from Caesar Hull, who was lost – together with Dick Reynell, ex-Hawker Test Pilot, from Tangmere on September 7th. It had taken eight days to get the tattered remains of their eleven Hurricanes from Tangmere to Usworth.

It was a shattered, but not downhearted, squadron that we joined. The veterans such as Frank Carey, Jim Hallowes, Killy Kilmartin, and Gorrie, welcomed us and set us about our business in double-quick time. I flew seventeen sorties in five days, starting with flight formation led by Frank Carey, then long-range local reconnaissance and sector exercise, followed by an intensive close formation and mutual dog-fighting exercise with George Westlake. We landed late to be refuelled and sent off on an operational patrol, our first. This was non-productive and after an hour we reported returning to base as it was getting dark, only to be instructed to divert to Thornaby as Usworth was 'clamped in' with low cloud and mist with nil visibility.

It was getting dark with failing visibility; Usworth had already alerted Thornaby and we called the latter to request landing facilities ... to be told they were closed and was our diversion necessary? We assured them it was and we were running short on fuel and repeated that Usworth was weather-bound and that they were wasting time and fuel as we orbited Thornaby.

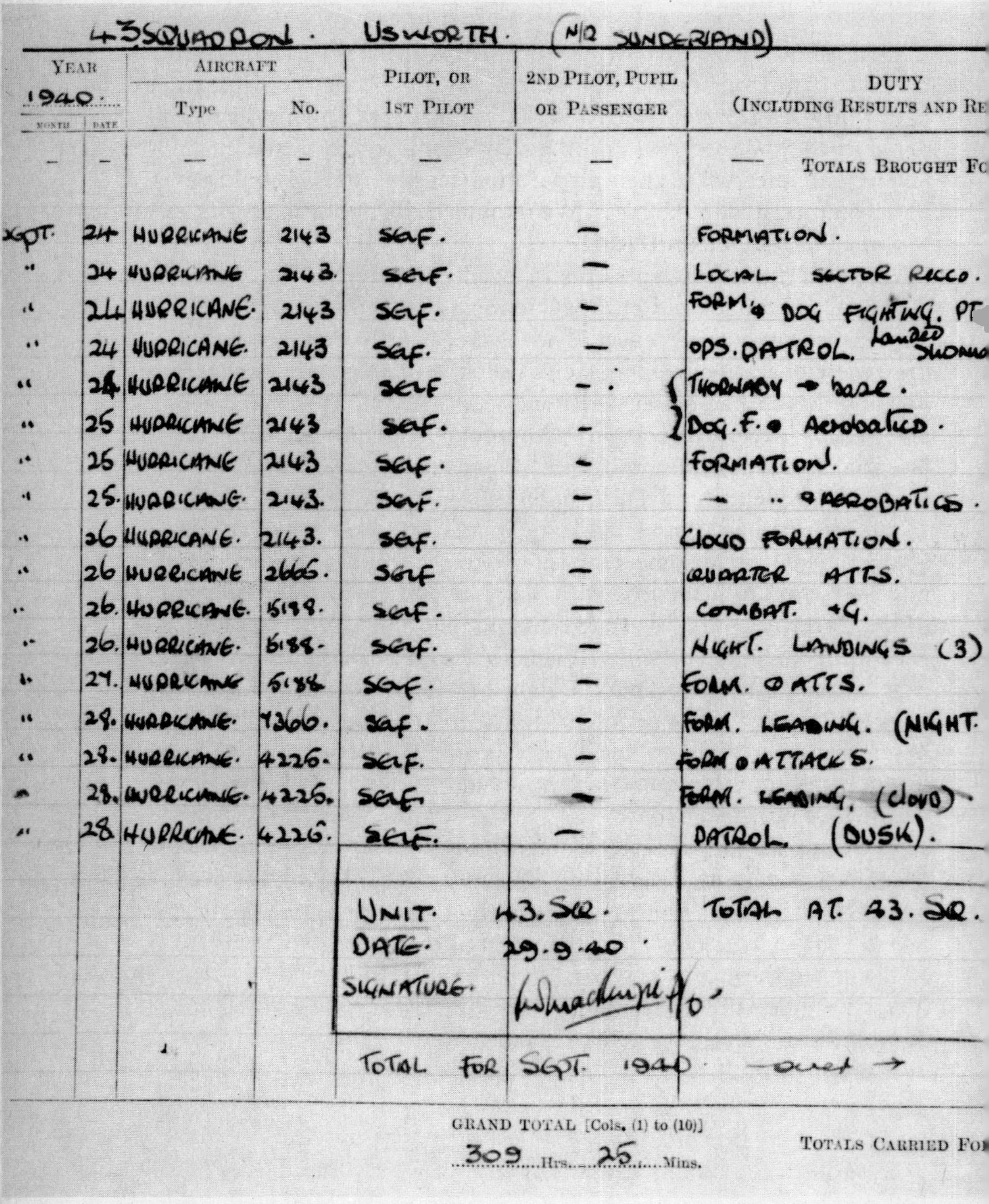

43 SQUADRON · USWORTH · (N/R SUNDERLAND)

YEAR 1940 MONTH	DATE	AIRCRAFT Type	AIRCRAFT No.	PILOT, OR 1ST PILOT	2ND PILOT, PUPIL OR PASSENGER	DUTY (INCLUDING RESULTS AND RE[MARKS])
—	—	—	—	—	—	— TOTALS BROUGHT FO[RWARD]
SEPT.	24	HURRICANE	2143	SELF.	—	FORMATION.
"	24	HURRICANE	2143.	SELF.	—	LOCAL SECTOR RECCO.
"	24	HURRICANE.	2143	SELF.	—	FORM. & DOG FIGHTING. PT
"	24	HURRICANE.	2143	SELF.	—	OPS. PATROL. LANDED [illegible]
"	24	HURRICANE	2143	SELF	—	THORNABY → BASE.
"	25	HURRICANE	2143	SELF.	—	DOG. F. & AEROBATICS.
"	25	HURRICANE	2143	SELF.	—	FORMATION.
"	25.	HURRICANE.	2143.	SELF.	—	" " & AEROBATICS.
"	26	HURRICANE.	2143.	SELF.	—	CLOUD FORMATION.
"	26	HURRICANE	2666.	SELF	—	QUARTER ATTS.
"	26.	HURRICANE.	5188.	SELF.	—	COMBAT. +G.
"	26.	HURRICANE.	5188.	SELF.	—	NIGHT. LANDINGS (3)
"	27.	HURRICANE	5188	SELF.	—	FORM. & ATTS.
"	28.	HURRICANE.	4366.	SELF.	—	FORM. LEADING. (NIGHT.
"	28.	HURRICANE.	4226.	SELF.	—	FORM & ATTACKS.
"	28.	HURRICANE.	4226.	SELF.	—	FORM. LEADING. (CLOUD).
"	28	HURRICANE.	4226.	SELF.	—	PATROL. (DUSK).

UNIT. 43. SQ.
DATE. 29.9.40.
SIGNATURE. [signature] F/O.

TOTAL AT. 43. SQ.

TOTAL FOR SEPT. 1940 — over →

GRAND TOTAL [Cols. (1) to (10)] 309 Hrs. 25 Mins.

TOTALS CARRIED FO[RWARD]

The author's log-book for five days in September 1940

After some further delay they responded and we landed and taxied in not in the best humour at their tardy attitude, later roused to downright anger. We were told that we had caused considerable panic as they were closed and there was a mess dance that night and we had interrupted their preparations; lastly, that we could not be accommodated in the mess. We reminded them that there was a war on in terms loud and clear.

We were met in the mess, us in battle dress flying kit, by the Station Commander and his PMC in mess kit, who expressed their displeasure at our presence! They were sharply reminded that there were operational emergencies which took precedence over his ball and that he had better take it up with the AOC as we had been diverted by Sector Operations. We apologised for disturbing his essential social activities and would he ensure that we had a room or rooms for the night ... Oh, the irony!

We returned to Usworth the next morning and reported the affair; the tale got us some free beers later! The following days were intensive training exercises with some of the old hands, Gorrie, Frank Carey and Killy, with two more night flights, ending on the 28th with a dusk patrol and being vectored onto a He111 over Newcastle. Visibility was poor with broken cloud but he was visible between layers; I got to about 400 yards and made a quick burst as he crossed from right to left, no strikes seen and no result as he disappeared into the murk; after a long patrol of one hour twenty minutes we returned to base.

The next day I was told to report to Number 501 Squadron (Auxiliary Air Force, County of Gloucester) at Kenley in Surrey; I was thrilled to bits ... down into the thick of it. George Westlake was posted to Tangmere, 213 Squadron. We made hasty farewells and thanked the 43 Squadron chaps for their 'blooding'. The short, intensive time spent with them had given us confidence in the aircraft and ourselves, good enough not to disgrace the new squadron.

CHAPTER SIX

Blooding

No 501 Squadron, commanded by Squadron Leader H.A.V. Hogan, a very senior squadron leader – efficient, cool, a good pilot and shot, for whom we had considerable respect as a leader – had recently been deployed to Kenley from Gravesend and Hawkinge. They shared Kenley with 253 Squadron, commanded by Squadron Leader G.R. Edge, DFC, lately moved down from Turnhouse in August. Together with No 605 Squadron (Auxiliary Air Force) based at Croydon, we constituted the Croydon Wing of Hurricanes and worked in conjunction with the Biggin Hill Wing of Spitfires with Nos 72 and 92 Squadrons at Biggin Hill and 66 Squadron at Gravesend.

No 501 Squadron had already seen considerable action with the Advanced Air Striking Force in France during the German advance and had only got out at the last moment via the Channel Islands. Reforming at Middle Wallop, they had immediately joined in the Battle, finally settling at Kenley. With an impressive record and high score, they still had many of the old hands such as 'Ginger' Lacey, 'Daff' Dafforn; Hawkeye Lee; Gus Holden; Morfill; Paul Farnes; to name but a few of the august company I was joining.

The squadron was made up of regulars and VR pilots, officers and sergeants, with a generous sprinkling of Poles who were irrepressible, irresponsible in some respects and a laugh a minute at times. We were glad of them as they had mostly received a sound blooding when their country was invaded and had made their way to England by devious ways and means to fight again. Their deep hatred of the Germans was infectious.

The ground crews and other personnel of 501 Squadron and the station at Kenley had all been in the thick of it for several months, the ground crews having been in France as well. These were again made up of regulars and Auxiliaries and VR, but by this time there was no distinction between them; all were equally involved in, possibly, a fight for survival and their cheerful efficiency and untiring efforts worked miracles to keep the station running, the aircraft serviceable, to give rapid refuelling and rearming for a quick turn-round. All this often savoured in the evenings, perhaps by impromptu relaxation in the messes or the local hostelries whenever time allowed, or when 'stood down'; for a brief rest or due to weather. These then were the squadron and station I was joining; truly a lot to live up to.

The Battle was entering the final stage. Three months or so had passed since the beginning of the, so-called, Battle of Britain and, although Fighter Command had taken a heavy toll and lost many aircraft, the Luftwaffe had also been badly mauled to the tune of some 1,700 aircraft in various ways. Germany had not established the air superiority forecast and necessary for the, soon to be aborted, invasion. There was considerable dissension amongst their own fighter pilots, who were tied to their escort role to protect the bombers and unable to exploit the somewhat better performance of the Messerschmitt 109 at altitude against our attacking fighters. Above 20,000 feet the Messerschmitt Bf109 was faster and had a better rate of climb than either the Hurricane or Spitfire, albeit they were not so manoeuvrable. This performance was however hampered and could not be exploited due to being restricted to flying, near, or just above the bombers that they were screening and protecting. They were restricted to inflexible formations and only broke when they or the bombers were attacked, inhibiting their fighting capability considerably, as the initiative was with the attacker. This factor was eliminated later in the final phase of the battle when the Germans introduced high flying fighter and fighter-bomber formations of Bf109's operating unfettered at higher altitudes.

Their daylight bombers were suffering unacceptable losses and new tactics were necessary. In late September the night blitz started in earnest, to pound British cities, whilst the daylight bombing attacks were accompanied by very large formations of Bf109's over relatively small bomber formations, together with the introduction of the fighter-bombers mentioned. This was of course all new to me, as we had previous but scanty detailed information at Sutton Bridge or Usworth, but the picture emerged quickly.

Arriving at Kenley was quite an experience. It was a full peace time RAF station with permanent buildings, good accommodation, very luxurious after my previous postings. My room was in the East Wing with ample room to park the MG 'PA' outside the back door, most convenient, good beds, clean sheets, a batman to look after one, good food ... well, hell, if it was going to be a short life but a hectic one, this was the way to live it!

I was immediately welcomed by the squadron and well briefed on what had gone before – the attacks on Kenley on 18th August and again on 1st September. The damage was extensive and visible. They were a likable bunch of officers and sergeants. Mostly experienced and very capable pilots, their spirit and morale were terrific, maybe forced at times after a full day – but genuine. They told me about the recent arrival of the Bf110's and 109's with bombs, the new tactics, spook raids and feints which were giving Fighter Command a headache. The heavy attacks made in the last few days, the 27th, 28th and 29th, when they were confronted over Kent by some 200 Bf109's escorting a single Gruppe of Ju88's, of the Germans staying to fight it out, tying down eighteen squadrons of Fighter Command and the three squadrons of the Croydon (our) Wing taking a pasting with the loss of several pilots and aircraft.

This was all tantalising stuff and almost too much to take in at once; obviously I had reached the action and was at once grateful for the Hurricane time at Sutton Bridge and with 43 Squadron. With a day to settle in, check my flying kit and draw a parachute, look round the airfield, the hangars – or what was left of them – get to know the ground crews and armourers, as I wanted to experiment with harmonisation if possible. It seemed that someone in 43 Squadron had given my new 'boss' a reasonable report upon me, because I seemed to be accepted immediately as one of them, which was really encouraging.

On October 1st I was told to do a sector recce (our call sign was *Mandril* Squadron), get to know the controllers and be vectored around a bit at low, medium and high altitude, but keep a good lookout for friendlies and 'bandits', come back and give us a 'spot of aerobatics over base!' What more could one ask for??

It was a quiet day round London, some broken cumulus cloud to disport around. I was vectored around the sector and at low altitude noted the local prominent features around the circuit area, the chalk pits in the Surrey Hills, the railway tunnel, the 'Purley Way', Biggin Hill across the road as it were. Whilst I was airborne I listened to the build up over the radio culminating in some 100 Bf110's and 109's attempting to bomb Portsmouth and Southampton and resisted an itching to go along and have a 'butchers' at what it was all about.

Returning to Kenley after over an hour's recce, I gave my authorised 'beat up' with a proper display sequence followed by a really tight circuit and spot landing. It was well received; 501 had not been called upon and our serviceability had risen in consequence. I was allocated Hurricane SD-X No V6799 just out of servicing and went over it with the fitters and armourers, not that I wanted to impress them or query anything, but to be sure that I knew my aircraft and got a 'feel' for it.

We were briefed in the afternoon about the introduction of a 'spotter' flight of Fighter Command being formed to fly high, avoid combat, and report the height and strength of approaching enemy formations as the speed of these fighter and fighter-bombers formations unencumbered by the heavy bombers was now so great that the radar chain could only give about twenty minutes' warning before they dropped their bombs. This did not give us time to 'scramble', gain height and attack the incoming enemy before they attacked. We doubted whether this tactic would in fact help; the only efficient way to meet these attacks was to put up standing patrols and we had neither the aircraft nor pilot strength to contemplate this. We heard that the first of the bomb-carrying Bf109's had been met, that Bob Doe of 238 Squadron swore that he had seen a Focke Wulf four-engined (Condor) in the bomber stream. This was later confirmed, I believe.

That night I listened to and watched the first night attack that I had witnessed against London. It was a sobering experience, starting around eight o'clock, with waves passing overhead to Liverpool, Manchester and, we heard later, to Swansea and Glasgow in very small numbers. I chatted to the squadron pilots and met some of the 253 pilots as well; their 'bar' chat was mostly shop, a good way of relieving the tension and strain, but a wonderful way for a new boy to pick up tips.

October 2nd started with a brilliant blue sky, clouding over later in the afternoon. 501 was stood down operationally, but I took my new SD-X for a short air test and to get the 'feel' of it with a really hard gut-rending work-out and find any snags ... there were none and 'she' felt good with a smooth engine giving full power. I liked it. In the gathering clouds I zoomed and dived and twisted and turned through the billowing valleys and clefts, full power dives and pull outs, maximum rate turns and climbs, tight turns and rolling through to turn the other way (later designated the 'Derry Turn' but used by us much earlier). Gauging the stiffening up of the controls from 250 knots upwards, the effect of trim and the marked tail heaviness in fast dives.

Returning to Kenley I learnt that again 501 had not been called upon, but that Biggin Hill had been attacked again and apparently the 'spotter flights' ordered by the C-in-C had proved their worth in alerting the defences of the several raids in the morning coming in at between 20,000 and 30,000 feet, effectively beyond the reach of our Hurricanes but eight or more had fallen to the Spitfire boys.

That night we watched, heard and felt continuous raids from about 7.20 p.m. to six o'clock the next morning. I frankly could not sleep and watched for hours, the impotence of our defences being highlighted as the fires were seen from our higher vantage point at Kenley, as London took a pasting. The searchlights did not reach the bombers and the AA fire seemed ineffective, whilst the Beaufighter night-fighters with their Airborne Radar (AI) had not yet become really efficient. London seemed helpless to defend itself and mercifully the German bombers did not carry a Lancaster's load! The Purley Way, the London-Eastbourne road, was a regular route in for raids on London and as we were only a mile or so west of it the distinctive drone of their engines became a regular night sound; in fact on quiet nights we missed them!

With little flying for us these few days I got to know my companions better: Vivian Snell, a humorous regular flying officer who was our tame Fats Waller on the ivory keys and kept us entertained at the mess piano; we had a mutual interest in fast cars, though his Morris 8 was not really in that category! Flight

Lieutenant Duke-Woolley of 253 Squadron was another enthusiast with his MG 'TA' which had been built and prepared for Le Mans by Monaco Engineering at Watford. Gus Holden, one of the flight commanders, a superb tennis player, who was well blooded in France, a good pilot and shot with experience as a regular from 1936. With an infectious sense of humour he was a good leader who did not waste words and seemed to be unruffled by setbacks.

His opposite number, a regular, was unimpressive and unreliable; we didn't miss his departure. Pilot Officer 'Daff' Dafforn, about six foot six inches of fighter pilot dominating the cockpit of a Hurricane; even with the seat down as far as possible he filled the canopy. A super guy, cheerful and reliable, another blooded in France. Hawk-eye Lee, another VR with a sallow complexion, a permanently drooping left eyelid – until he sighted an enemy aircraft – or a bird. 'Stan' Skalski, a Pole, slightly built, pale-faced with an emotional hatred of the Germans, together with his compatriot Kozlowski, a quiet, unassuming pilot who never showed his feelings.

The sergeants were a mixed bag, mostly VR, headed by Ginger Lacey, who did well in France and ended the Battle of Britain as top scorer. A slight, almost ginger-haired boyish figure, you would think that butter would not melt in his mouth ... and how wrong you would be. Well-blooded in France, he had joined the VR with Hawkeye Lee whilst flying as an assistant instructor to the Yorkshire Aero Club and was therefore an experienced pilot when war broke out. He used his experience to good effect as results were to prove. He was direct, intolerant and had a marked indifference to others with a macabre sense of humour; we shared a mutual dislike of bull and mock heroics and a love of flying to the limit.

There was Paul Farnes, quiet, meticulous and unassuming; a very capable pilot and 'Morf' Morfill, a regular, taciturn, quiet, capable, reliable; both experienced from France: Tony Glowacki, a Polish sergeant, squat, tough, always laughing no matter what, already experienced from the German invasion of Poland, a smooth pilot and good shot. Tony Whitehouse, another sallow-skinned smooth operator, with the flashiest car, an 'SS' Saloon, on the squadron, and a wicked sense of sardonic humour. Vic Ekins or 'Farmer Giles', stolid, reliable, steady as a rock. Konrad, a Pole who looked more like an Italian, quick and voluble until his English ran out, and many more who made up the squadron strength.

I got to know and like them all and others like them. There was one thing that stuck out a mile – their mutual respect for one another as pilots, fighter pilots and as men, the total absence of in-play, criticism or jealousies. The team spirit was there in abundance, but the individualism of each fighter pilot stood out, and the standard of flying was second to none except for one or two who inevitably fell by the wayside.

I had joined a formidable band of warriors indeed. The ground crews, too numerous to pick out, were real characters in their own right, the backbone of the squadron. We could not let them down, no matter what.

This was the scene on October 3rd when we woke to low cloud and drizzle covering most of England with Fighter Command virtually grounded. We were stood down again but later called to readiness as sneak raiders penetrated using cloud cover and coming in low to fox the radar. We learnt that a Ju88 had bombed the De Havilland works at Hatfield with many casualties and destroying much of the materials ready for the magnificent new Mosquito aircraft; a serious setback for them to overcome.

Stood down after lunch, as visibility was by now nil, we took the opportunity to examine the many derelict motor cars that had been dispersed round the airfield to use as obstacles if the Germans ever tried to land gliders or troop carriers. They were a motley collection of junk, but two were usable, an 'L' type MG tourer and an Overland saloon with American split-rimmed wheels, a large 'gharry' and eminently suitable as squadron transport, naturally neither taxed nor insured! This, with some help from the 'MT' boys, we got running, and took that evening on a quick mission to the Surrey Hills Hotel, the bare metal rims clattering along the road in grand style!. With six up, the return journey up the Caterham hill was a tour de force with the metal rims sending out a shower of sparks. Our arrival at the main entrance and the guard room was heralded well in advance and merited some choice words from the guard commander. Our arrival at the mess resulted in quite a party ... and so to bed.

October 4th brought my first fully operational flight with 501 Squadron, first flying as Blue 2 in a flight of four on a reconnaissance sortie which produced nothing, in very bad weather, mist and rain in patches, with difficulty in finding Kenley on our return.

'Scrambles' were rapid scrambling into parachutes and into the cockpits of our aircraft to take off, hence the name. They were usually from various 'states of readiness'. Immediate, fifteen minutes or thirty minutes call when we would be at squadron strength at dispersal waiting for a call to take off and climb to a designated height on a predicted course from the fighter controller to intercept, or attack a raid or join other squadrons to do so.

The aircraft were dispersed some distance apart, or in blast proof bays with starter batteries connected and our parachutes hung on the port wings ready to strap on as the ground crew positioned themselves for the start-up and released the chocks under the wheels. Some pilots liked to keep their parachutes in the cockpits and to jump in and strap up *in situ*; I and others, found the cramped space fiddling and it was difficult to get sat on the parachute and strapped in satisfactorily and rapidly at once. Pulling it off the wing by the top harness, the whole 'chute swung down into place and it was quick and easy to do up the bottom harness, climb up onto the wing, into the cockpit, do up the cockpit harness aided by the fitter ... into place in one!

Pilots usually sat round their dispersal huts, or inside in rain, nattering, playing cards, making model aircraft or what have you, ready to drop everything and run when the scramble was called. Such as: 'Mandril Squadron, scramble, Angels 20, course 110', for example.

The take-off was as best possible into wind and sometimes not, with several aircraft at once, in pairs, threes, ones, at 'full chat', settling into the cockpits, adjusting flying kit, seat height, checking instruments as we did so, watching each other as we were very close. A hectic few minutes packed with action as we formed up in our respective flights and climbed away on the course given, establishing contact with Control and our Identification Friend or Foe (IFF) link which identified us on the radar to avoid confusion and save time identifying through other means which would give our positions away.

If there was time in hand we would form up over base, or if we had to penetrate cloud, but when time was critical the leaders would throttle back so that all could catch up on the climb out. Radio contact between all aircraft would be checked and all pilots would be 'listening out', and we endeavoured to strictly restrict chatter to essential communication only. Once in formation, of whatever type, the climb would be regulated by the controller who was interpreting the overall picture and could be a rapid climb to height for positioning later, or a slower climb out towards the target combining distance and height.

Throughout we would be controlled by the sector controller and vectored as necessary towards our quarry, or positioned advantageously to intercept in relation to other squadrons and the enemy position and numbers, with a running commentary of information being given. This information was obtained from the radar plots and the Observer Corps and listening posts in the Sector; fed into the Operations Centre (Room) and correlated by the controllers. Once in position it was a case of weaving, being alert, watching, looking for the enemy if not immediately visible, covering our rear, keyed up and ready to attack, or break into an attack if jumped by the enemy fighter escorts.

When battle was joined, it was every man for himself really, although we did try to operate in pairs, one covering and backing up the other. This was wasteful when attacking bombers and we tried to single out separate targets initially, but as soon as the attack developed the sky was a milling mass of weaving, shooting aircraft and it was fortuitous if one was warned of an escort fighter latching on to one's tail.

The action was rapid, intense and short-lived. In a matter of minutes one could suddenly be alone in an empty sky and wonder what the hell it was all about, where had they all gone, or had there been several hundred aircraft there at all? Then came a quick appraisal of the situation, listen to the radio chatter, look for more targets if one still had ammunition, rejoin the squadron or join up with one of them, but above all never fly straight and level in the combat zone; watch your tail, weave, search for the battle or indeed break off if out of ammunition and make your way back to base. Sometimes we were able to rejoin in squadron strength, or flights, and fly back to base in formation, more often we returned in ones, twos and threes ... some never returned, some baled out or force landed to return later by road.

Upon returning to base we would land, taxi in, get out and inspect the aircraft for any damage, report any technical snags to the ground crew for attention, leave our helmets and parachutes with the aircraft and report to the Intelligence Officer who would be waiting eagerly to debrief us. The debriefing was short and sharp, usually one at a time with others listening who could back it up, corroborate or dispute the action described, help to cross-check the action and results. We made out separate combat reports if there was time before another sortie. The Intelligence Officer would interpolate the reports when completed and pass the initial information to Sector Operations. Claims would later be related to enemy aircraft found destroyed, reported in the sea, watched by onlookers etc., aircraft known to be shot down or damaged as claims were assessed. From this the attrition rate to the enemy could be calculated pending confirmation and further information. Our own losses were similarly interpreted.

With the numbers of enemy and own aircraft involved in a relatively small column of airspace in a very short time, it was inevitable that initial claims could be exaggerated, duplicated and confused as several pilots shot at the same formations, attacked the same aircraft, or one already damaged, saw the same aircraft go down, burst into flames or be damaged, or the pilot bale out. The action was intense, hard physical work with mental alertness the keynote, but wishful thinking and blurred vision could sometimes cloud observation of the facts under such stress. We could not possibly follow every aircraft we shot at, or damaged, or indeed should have unless it was to complete a 'kill': to do so would have fragmented the attacks and wasted time and effort. In the battle one would attack, may be forced to break off and attack another, or take evasive action to avoid an attacker oneself. We endeavoured, as far as possible, to report accurately the position of enemy aircraft destroyed or damaged, but mistakes were made on both sides. In the main our claims were more conservative than the Germans, as subsequent investigation proved.

This pattern of operations I soon accepted, lived with it and enjoyed it to the full. To me it became a challenge, impersonal, machine against machine with the great satisfaction of being as one with such a superb fighter: the Hurricane.

This then was the scene when, on October 4th, we were scrambled for the second time at around 11.00 hours. The weather was still mist and rain with drifting cloud in layers. Enemy activity was fighter-bomber and fighter Bf109's using cloud cover to penetrate to London in an almost continuous stream.

We were in four sections of three aircraft and vectored in broken cloud to intercept Ju88's over Kent in the vicinity of Folkestone. The weather seemed against us and maintaining formation took most of our concentration. Skirting cloud at about 8,000 feet, control directed us to look out for a Ju88 crossing our flight line at 1,000 feet below us, as we vectored onto a course of 100 degrees. Control told us he had turned towards us about three miles ahead.

Blue 1, Flight Lieutenant Holden, and I, spotted the 88 ahead and below coming towards us almost head on. We both 'Tallyhoed', Blue 1 called 'attack and gate'. The other section stayed up and on course covering us.

We eased on full power, boost and revs up, Gyro sights 'on', guns to 'fire' and winged over to attack. Blue 3 fell behind for some reason and I flew close formation on Blue 1 expecting him to attack head on, but he turned away and then pulled round to deliver a full deflection shot from the stern quarter, opening fire at about 300 to 350 yards: no hits seen.

I followed him round with a beam attack, cutting off the corner in a very tight turn with a closing speed of about 80 to 100 knots. This developed into a stern attack very close to the enemy and I opened fire at about eighty yards with a long burst into the port engine. Smoke and flames came from it and some bits fell off it.

The rear gunner then had me in his sights and opened fire hitting my port wing, but stopped abruptly as though he had been hit, perhaps by Blue 1 in his initial attack. Blue 1 then joined me as I broke off the attack overflying the 88 as he entered some wispy cloud. We followed him through broken cloud as he turned down the coast towards Eastbourne, the port engine belching smoke but no longer on fire, as he lost height in a shallow dive. We attacked again but lost him as he weaved through the cloud, a pity as we had such a speed advantage and he was obviously in trouble.

Blue 1 kept above but I had to break off the chase as we were now in thick cloud and visibility was a few hundred yards only. Blue 1 and I made radio contact and met up at about 8,000 feet where the cloud was less, joined up and made our way back to Kenley, which we found with considerable difficulty by following the Purley Way (main road) and 'hoisting' ourselves up onto the airfield, with visibility of about three-quarters of a mile.

I was tickled pink, second operational sortie and one Ju88 claimed damaged, shared maybe. However we had had two aircraft from the other flights forced-landed due to the weather – Flight Lieutenant Jones and Ginger Lacey; they returned the next day. From the German records it was later confirmed that a Ju88 crash-landed (No 8076) at Abbeville around the same timescale 'after combat with unknown British fighters'. So, perhaps a 'confirmed' could be justified as it was a 'write-off' with one crew member wounded ... the rear gunner perhaps??

That evening we celebrated, I had been blooded, to wake with a somewhat sore head at an early call to readiness next morning at six o'clock ... self-inflicted of course!

CHAPTER SEVEN

Really Into Battle

October brought in the next phase of the Battle. We now knew that Fighter Command losses for August and September had been some 715 fighters 'Cat 3' (Reduced to produce) in combat and some 307 pilots killed or missing, with 308 wounded. The number of replacement fighters supplied was 302 Hurricanes and 319 Spitfires, or rather less than the number lost. Experienced pilots were in very short supply and new pilots had less and less hours on posting to operational squadrons. This knowledge made me appreciate my time on Hurricanes at Sutton Bridge and with 43 Squadron more than ever.

Next morning at dispersal the CO, Squadron Leader Hogan, met me and said, 'Congratulations, Mac, you're a mad bugger ... glad to have you with us!' From him, that was praise indeed and it made my day!

I was still concerned regarding the 303 guns and again tried to get the 'Chiefy' Armourer to reharmonise mine down to a two foot diameter circle at 200 yards, I was convinced that had they been so sighted we could have clobbered that 88 once and for all instead of nibbling at undoubtedly very robust engines with 'buck shot'. The Ju88 was a notoriously strong aircraft and difficult to shoot down, certainly if not from close range. Though I had got the port engine from about 80 yards, my burst of fire was too short owing to my high closing speed and I should have throttled back to be able to get in a sustained burst, I made a mental note for next time. In discussion, we all agreed that the best way to get an 88 was to clobber the front crew compartment where all the crew were, if one could do so, either head on, or from astern above, though of

ıe author marshalling at
ewtownards Airport, N.
eland, during the 1937 King's
up Air Race, seeing in Waight
the De Havilland TK4.

ex Henshaw refuelling at
ewtownards Airport, N.
eland in the 1937 King's Cup
r race, flying the Miles
awk'.

aff (Flight Lieutenants Sloan
d Lindsay and Flight
ergeant Lake) with trainee
ots and navigators of No 4
&R FTS, Northern Ireland at
denham airfield Belfast.
pril 1939. Author third from
ft.

Author in Hawker Hart, 1939. Sydenham, Belfast.

DH Tiger Moth – one of the basic trainers for all RAF pilots during WW2.

Miles Magister.

Flight-Lieutenant H.J.L. Hallowes, DFC, DFM. 43 Sqdn.

Wings over Westminster – combat contrails over Parliament, Summer 1940.

Hurricanes I's October 1940.

course a good deflection shot would be as effective but more difficult.

October 5th followed a disturbed night, with a constant stream of droning raiders en route to London – 200 or so – some of which had split off and attacked Liverpool, the London attacks being against armament factories using large bombs and land mines, most of which fell on residential property. We were called to 'readiness' at seven o'clock and kept at '15 minutes call' until 8.30 when we were brought to 'immediate readiness'.

The weather was better, light rain with bright patches, ideal for sneak raiders, unfortunately, and the enemy was well aware of this. This was to be the curtain raiser in earnest for the new fighter-bomber high altitude raids and the first build-up began with 30 plus off the coast at about nine o'clock, with more to follow in forties, thirties and twenties, down to as few as twelve or so aircraft, difficult targets, high and fast. Mostly they were Bf109's but there were several formations of Bf110's, some as bombers too. It was really boiling up and a full scale effort by Fighter Command was in the offing. With a complete squadron of aircraft and pilots available after a few days of 'bad weather servicing' we felt ready to go.

At ten o'clock we were scrambled in squadron strength to patrol at 28,000 feet over the Kent approaches, a sustained patrol listening to other interceptions and controllers' instructions, we saw distant dog fights but were not called in. As we weaved and looked out, the cold after an hour was intense without cockpit heating. It was necessary to sit on one hand at a time to keep the fingers mobile and there was some icing of condensation on the inside of the windscreens; this restricted forward visibility and we had to constantly try to keep a clear spot to see through. For this we carried a short stick with a chamois ball on one end soaked in glycol. I kept mine on a piece of string so that I couldn't drop it into the inaccessible nether regions of the cockpit.

We found 'no trade' and were not vectored onto anything. An abortive sortie but good to get the full squadron up again. Landing at 11.30, we had a quick refuel and went back on readiness. Scrambled again at 11.45, we were instructed to climb to 20,000 feet but before reaching 18,000 we were vectored to intercept a formation of Bf110's and 109's given as 30-plus at 16,000 feet.

Climbing to gain some more height we 'bustered' with the sun behind us and spotted the formations. With the tit pulled for absolute full power we broke formation to attack, Gyro sights 'on', guns to 'fire', harness tight, attacking individually. They had not seen us and we had a superb advantage, they broke wildly on seeing us and being shot at, I took a Bf110 very fast, too fast, from astern and below with a four second burst from 250 yards, broke away to avoid another 110 but saw no apparent damage to the one attacked.

As I pulled away and up to attack, a Bf109 above me, another Bf109 flew across my flight path, so I reefed hard round nearly blacking out to get a full deflection shot at him from 100 yards. Ah, that was better! ... Strikes visible all over him but our crossing speed too fast. He slowed visibly as I pulled up and above him as he dived. With my height and speed advantage, he did not draw away, as the 109's usually did initially, and took no evasive action as I followed him at about 1,200 yards and closing, no other aircraft about – we had lost them, though I kept a wary eye on my rear view mirror. Puffs of smoke and then white smoke came from his exhaust as we neared Margate, going south-east. When just off Margate I followed him at 600 yards down over the sea, closing fast, as he levelled off at about twenty feet – still no evasive action. I closed to 100 yards and gave him a long burst from dead astern, watched the strikes, but no immediate result. Throttling back I waited and suddenly he just dived into the sea in a cloud of spray so that I had to pull up sharply to avoid his tail as it came up.

I zoomed up to 2,000 feet, had a good look round and orbited the spot, but there was nothing to be seen but oil and disturbed water. After a couple of orbits, I returned to base at about 12.30 to find that the CO, Squadron Leader Hogan, and Flight Lieutenant E. Holden, DFC, (Gus) had each got a Bf109. The latter had seen mine go down with me chasing it as he had shot his down, which was apparently about to attack me ... I thanked him!

We had a quick lunch and came back to readiness again at 14.30 with another scramble at 15.00. This time, together with other squadrons, we were vectored at 26,000 feet to intercept a raid of fifty-plus fighters and fighter-bombers over Kent on the Ashford-Tenterden line. Although well placed with a height advantage we were late in finding them and arrived after the other squadrons had broken them up and driven them off with some losses, so no trade unfortunately and we returned to be stood down. It had been a tiring day and the cold at high altitude was taking its toll, whilst the strain of higher speeds and longer flights was affecting some. It was a relief when we heard that we were stood down until the 7th.

That night London took another pasting, but we slept through it after a relaxed evening. The following day, October 6th, was, as forecast, wet and cloud-covered with continuous rain. Enemy activity was reported as slight by single or small groups of aircraft against our airfields. Biggin Hill was bombed by a single intrepid raider with three barrack blocks destroyed. Our sister squadron, 253, was scrambled and Flight Lieutenant Duke-Woolley and Murch shared a Dornier 17Z, a type not often seen these days, except in bad weather. Northolt was attacked and a Hurricane destroyed, killing the pilot whilst taxiing.

We used to stand down to service cars, do some local shopping, play squash and I also spent some time with the armourers about harmonisation, as the debriefing after the squadron attack on the previous day had highlighted the necessity to get more weight of lead into the enemy and pilots down to closer range. I had an idea about making a sort of gunnery trainer to use on the ground which the armourers were a bit sceptical about initially. I was by now being chaffed about my fixation on getting in closer.

That evening we betook ourselves almost in squadron strength to the Greyhound Inn in East Croydon, where the manager was very 'pro RAF' and the steaks were like baby calves, the customers welcoming and jolly with a good pianist and sing-songs ... bawdy and otherwise ... the head waiter, 'Steamboat' I think we called him, treated us right royally. Indeed sometimes when it came to paying we would find that someone had met the bill for us, a most kind gesture but embarrassing. The convoy of battered vehicles returned safely, the Overland being passed by the police with never a glance, although it was neither taxed nor insured!

Notably our CO Squadron Leader Hogan, and Flight Lieutenant Holden, OC 'E' Flight, both somewhat older than most of us, did not join in these evening excursions and I think that they looked upon them as being rather infra dig. Maybe so, but it did ease the tensions and most of the pilots were the better for it. Regarding physical fitness, not much was doing in this respect during these days of high intensity flying and indeed the effort put into several sorties per day was the equivalent of several hours' hard exercise and we seemed to keep remarkably fit this way alone.

Although sharing Kenley with 253 Squadron, it was noticeable how the two squadrons kept their identity in the messes and on a personal level we never really got to know many of them well.

The next morning, October 7th, dawned with better weather, reasonable visibility, some showers but obviously a busy day ahead. We heard that parachute mines had been dropped the previous day at Uxbridge, headquarters of our most important 11 Group. One of them, a delayed action type, was apparently defused by a Naval sub-lieutenant, earning him the George Cross.

Brought to readiness early, we were scrambled around 11.00 against a large force but missed the main attack and had no trade, landing back after a very long, 1.30 hour sortie at maximum endurance with the weather clearing. After a rapid refuel to 'stand by' we were again scrambled with other squadrons against a reported 150-plus in waves from Calais to Dover at different flight levels. It was touch and go and 11 Group had to call for reinforcements from 12 Group. This time there were Ju88's mixed in with the waves of Bf110's and 109's, a formidable force, but capably met by our group commander and the controllers.

We took off in squadron strength, Squadron Leader Hogan leading to patrol the Sevenoaks approaches in company with No 605 Squadron, at 20,000 feet. First we were vectored south, then east and whilst flying east the CO saw two Bf109's flying in front of us and below, going south-east exactly as informed by Control. These two split up as we broke to attack them and another larger formation which appeared below us to the east.

In the ensuing dog fights I was attacked and got clear and followed the CO down as he attacked a Bf109 and scored some hits in a long burst, Red 1 (the CO) broke away to the right as the Bf109 entered cloud, I broke above the cloud at about 6,000 to 6,500 feet, being pretty sure that the Bf109 would come out on a reciprocal course, or certainly not straight ahead! He did in fact come out at right angles and then turned towards the sea, with a superb height advantage I half rolled after him giving the Hurricane everything it had got, closing fast from astern I hit him fair and square in the radiator and under the engine from 150 yards, he slowed rapidly, lost height and ditched off Hythe.

I watched the aircraft sink slowly and the pilot get out.* With considerable activity still visible high above me, and with some ammunition left, I climbed quickly to 23,000 feet, hearing the squadron still dog-fighting nearby and the controller reporting the enemy positions I decided to patrol between Dover and Folkestone at this height, keeping a good watch on my own tail. At about 13.30 I spotted eight Bf109's coming across the coast from the east about 1,000 feet above me. With ample speed in hand I pulled up into a loop under them and half rolled off onto the last three when a fourth came into sight a little ahead and slightly above me; obviously they had not seen me, or so it seemed. He was so close I gave him a quick burst in the belly from about 200 yards, noting many strikes around the radiator and under the cockpit.

He took violent evasive action and half rolled into a dive and nearly hit me in doing so. It took me a few seconds to recover and dive after him, by which time he had drawn away to about 800 or 900 yards with glycol streaming from his radiator. I pulled maximum boost and began to catch him as he shallow-dived towards Folkestone.

Passing close to the harbour I closed right up and opened fire at less than 100 yards; I could not see any direct result and my guns went silent after two or three seconds. What to do?? He was down to about 180 knots and losing height flying straight and level, no evasive action whatever as we got to within about 80 to 100 feet of the sea, watched as I subsequently found out by AA gunners and others round the Folkestone harbour.

At this stage I was determined that he should not get away as the pilot would undoubtedly be picked up by their efficient rescue service and live to fight another day. Having got him to this state he

*This aircraft was the subject of a dramatic recovery 35 years later by the enthusiasts of the Brenzett Aeronautical Museum and the book *The Reluctant Messerschmitt* by Don Everson.

had to be finished off somehow and quickly before we were too close to France. Indeed it was like doing close formation practice, with a difference!

I flew alongside him, round him, signalled him to ditch; he never looked up but seemed slumped over the control column. I lowered my undercarriage thinking to knock his tail off with it, but it slowed me down so much that I lost ground. Raising it I formated on him on the port side and in an impulse smacked my starboard wing tip down onto his port tail plane, it broke off, I pulled up to the left as he plunged in and about three or four feet of my wing tip flew up into the air, he partially sank without opening the canopy.

I did not wait to see any more as two more Bf109's were attacking me from above and behind ... they kept a protective patrol over the Channel to cover their air-sea rescue people ... I shot down to within a few feet of the sea, jinking madly to left and right, up and down. With no ammunition left I was a sitting duck and they were a persistent pair but not very good shots and mercifully did not seem to be armed with cannon, but their machine guns were enough and they hit me in the radiator and the armour plate behind my seat. As I approached Folkestone again they broke off and I managed to pull up and crash land on the hill to the immediate north of Folkestone, not far from Hawkinge aerodrome, undercarriage up without too much damage to the aircraft. It was subsequently repaired and put back in service.

In the forced landing I had, foolishly, undone my cockpit harness for a quick exit on landing, a grave mistake, as on landing I was jerked forward sharply onto the gun sight, splitting my jaw and smashing some front teeth.

Making a hasty exit I was met by some chaps from a nearby Army gun post who had witnessed the whole chase, out and back, downing the Bf109 into the sea and the attack and my escape from the two Bf109's and then nearly landing on top of them.

In a short while I was picked up in an Army car and taken to a Naval sick quarters in the harbour area and examined by a Naval doctor as my bottom jaw was cut open across my mouth; in fact I found that I could put my tongue out through it!! He was a grand fellow but had been, bless his heart, called from a lunchtime session and was, frankly, nicely, thank you ...

No matter, all was well, he seated me on his operating couch with my back against the wall as he approached me with a large curved needle threaded with gut and a wicked gleam in his eye. I watched the wobbly hand with great trepidation, but with a magnificent jab, steady as a rock, he pierced my chin four or five times in quick succession ... three plain and two purl, I thought ... In a flash he had done it, the gaping hole secured and it is a beautiful piece of stitching to this day. The local Commander of the Defences sent a message asking me to tea; this kind offer I reluctantly declined; my mouth was in no shape for food or drink!

The sequel to this episode was to come in 1948 when posted to the Air Staff at Headquarters Fighter Command, at Stanmore, after the war as a squadron leader and we were moving into a house at Northwood and had Pickfords 'do' the furniture. Coming back at lunchtime one day to see how the moving-in was going, I spoke to the chaps unloading and went into the house. One of them went to my wife and asked her – I was, of course, in uniform – 'Did he ever land a Hurricane on its belly near Folkestone during the Battle of Britain'? 'Yes', she said. 'Why?' 'Well', he said 'I was one of the gunners on the gun post that he nearly landed on, who went to help him out of the aircraft after he crash landed and as we approached he was already out of it and he pulled us down onto the grass, as he thought the aircraft might go on fire. But it didn't and we helped him over to our tent near the guns. He had a large gash in his jaw!'

Naturally, this called for the usual 'cuppa' and we sat down and chatted about it. I could not recall the bit about the aircraft going on fire, though. A small world and nice to meet him again in better times. We asked him to call any time he was in the area but I never saw him again and wished I had got his address, or even his name written down.

After the Doc had sewn me up, I waited to be collected and driven back to Kenley. Another mild celebration was held that evening as it seemed that, together with No 605 Squadron, we had had rather a field day, but lost Flying Officer Barry, killed. 601 Squadron also had a lively day. Our squadron score was three confirmed, but two of our aircraft lost, one destroyed and the other damaged. We heard that Westlands aircraft factory at Yeovil had been bombed. It had been a busy day.

The following day the station dentist examined my front teeth and, although I had a plaster on my jaw over the stitched wound, he diagnosed pyorrhoea in four front teeth and then asked, 'With such bad teeth, how on earth did you get selected for aircrew?' I nearly kicked him in the teeth then and there, but instead told him in a few succulent phrases! Life can be difficult at times, I thought.

Leaving him, I sought out the CO and told him the problem, that at least four teeth would have to come out and could I get over to Belfast quickly to a dentist friend, not trusting our local lad. As I was off flying anyway due to the wound, he readily agreed and offered me the squadron hack, a beautiful old Hawker Hind – my first love – to fly over in, but I went by train and boat instead.

Arriving in Belfast, I contacted Joe Lowe, a well-known dentist and friend of the family, whose practice was adjacent to the notorious Crumlin Road jail! He fixed me up immediately, relieved me of four front teeth fractured in the crash landing and supplied a new set in a couple of days. Service with a smile in good Ulster fashion. It was good to be back, but not for long. I was fortunate enough to be awarded the Distinguished Flying Cross whilst on 'dental leave' and an appropriate celebration was held before I returned to rejoin 501 at Kenley on 17th October, ready for anything.

I heard about the squadron's activities during my absence: the latest enemy tactics of sending covering fighters in at 30,000 to 32,000 feet who were inflicting increasing casualties on our aircraft intercepting raids at 24,000 to 28,000 feet, the Hurricanes at nearly their maximum ceiling, though the Spitfires were still reasonably well placed. These very high-flying Bf109's had a terrific height advantage and were a dangerous threat. 501 had not suffered badly; 253 had lost a couple of aircraft but the pilots were OK; Ginger Lacey had got another 109 into the sea on the 12th and claimed that a Hurricane had shot at him on the 15th – unsuccessfully as it transpired; he was not hit!

Flight Lieutenant E. Holden on the 12th, flying at 23,000 feet, attacked a Bf109 and chased it down to ground level and finished it off with a long burst which produced 'volumes of black oily smoke from the engine which entirely covered his windscreen as the enemy aircraft dived into the sea off Dungeness'! watched by 'Ginger'

Lacey! In the same engagement the CO Squadron Leader Hogan, got another Bf109 from 23,000 feet with several short bursts at 100 yards, later seen to crash south-west of Rye.

This was all that I had missed; I was frankly envious. It was also very interesting to note that the 'killers' were getting in to close range whilst several others who made good attacks on enemy aircraft but opened fire at long range, were just not getting the 'kills'.

I had lost my faithful Hurricane SD-X, V6799 and would have to take anything available until another came along, but not yet. That night Kenley was attacked during raids over Kent and London, four aircraft were damaged by bomb splinters and we experienced the effects of a direct raid for the first time at night. It was frankly unnerving to all of us, who were used to being able to fight back immediately, with the frustrating feeling of impotence, of being defenceless, hearing but not seeing either the aircraft or the bombs.

It made us talk about the apparent cleanliness, as it were, of the fighter pilots' environment at this stage of the war, the comparison, or analogy, being of Knights of Old jousting on trusty steeds; short; sharp, intense, followed by good living. The bombing was dirty, sneaky, prolonged, nerve-racking; attacking civilians, children, old people, destroying a way of life. No individual combat, no chivalry; not just the destruction of man and machine, but of property, livelihood, indiscriminately against those who really had no part in the conflict by design or choice.

It was a salutary experience which seemed to affect us all deeply for, from that day on, life took a more serious aspect. We were never again as jocular or light-hearted. Maybe it was recapping of previous months of intensive action, strain, but a sudden maturity enveloped us in 501 and 253 Squadron seemed to be similarly affected.

The next few days were some of the quietest, with limited visibility. The CO briefed us on the problems facing the 11 Group Commander, Air Vice Marshal Keith Park, and of his new plans to 'get at' the high flying fighter-bomber and fighter raids. These were as follows:

Engagement of High Fighter Raids

The general plan is to get one or two Spitfire squadrons to engage enemy fighters from above about mid-Kent, in order to cover other Spitfire and Hurricane squadrons whilst climbing to operating height at back control lines east and south of London.

Preparation

Whenever the cloud conditions are favourable for high raids by fighters the following preparations will be made:

(1) Reconnaissance Aircraft: One or two reconnaissance aircraft to be kept on patrol near the Kentish coast, height depending on cloud layers.

(2) Readiness Patrol: A patrol by one or two squadrons to be maintained on Maidstone line at 15,000 feet, between 0800 hours and 1800 hours.

(3) Stand-by Squadron: One squadron at sector providing patrol at (2) to be standing-by during the peak periods – breakfast, noon and early tea-time.

(4) London and Debden Squadrons: State of readiness of Hurricane squadrons to be *advanced state* whenever cloud conditions are suitable for very high fighter raids.

Attack

Immediately enemy formations are plotted over the French coast or Dover Straits, the following action will be initiated:

(1) Reconnaissance Aircraft: Despatched to the area enemy raids are plotted, to locate, shadow and report.

(2) Readiness Patrol: Ordered to climb to 30,000 on the Maidstone patrol line to cover other squadrons whilst climbing over base patrol lines.

(3) Stand-by Squadrons: Despatched to rendezvous over base at 20,000 to 27,000 feet, and when assembled, detailed to raids or forward patrol lines.

(5)Squadrons at Available, Spitfires: To be brought to readiness, and if necessary despatched to assemble in pairs on back patrol lines at 25,000 to 30,000 feet, and then detailed to raids.

(6)Squadrons at Available, Hurricanes: Brought to readiness, and if there is a second or third wave, assembled in pairs over back patrol lines so as to protect sector aerodromes and London area whilst climbing.

(7) Hurricane Squadrons from Tangmere and Debden: Despatch in wings or pairs at 20,000, to 27,000 feet according to time and weather conditions, of one of the following purposes:

(a) To reinforce London sectors if there is a second or third wave of enemy raids;

(b) To protect sector aerodromes and London area whilst the earlier squadrons are refuelling.

(8) Close Defence of Important Bombing Objectives: If enemy raids are approaching aircraft factories, London area, sector aerodromes, etc., single Hurricane squadrons that have not been included in pairs or wings should be detailed to protective patrols between 15,000 and 18,000 feet depending on clouds.

He explained that the Group Commander was concerned that the enemy might revert to their massed bomber raids and that our tactics were to be ...

Spitfire Squadrons:
Assembled at height in pairs on the back patrol lines, then detailed to engage high fighter screen at 30,000 feet.
Role: To protect pairs or wings of Hurricane squadrons whilst climbing up, also while attacking bombers plus escort. If the high fighter screen withdraws to the coast a proportion of the Spitfires may be detailed to attack the escorts to incoming bomb raids.
Hurricane Squadrons:
Squadrons at readiness to be despatched in pairs to back patrol lines covering line of aerodromes. Immediately pairs have reached operating height, detail to bomb raids or to forward patrol lines under Spitfires. Squadrons at available to be brought to readiness and assembled in pairs at operating height on back patrol lines covering sector aerodromes, and detailed to second wave of bomb raids. Whilst gaining height the latter squadrons may have to be detailed to split raids by bombers that attempt to attack vital points on the flank of the mess of bombers plus escort.
Hurricane Squadrons from Flank Sectors (Debden, Tangmere, and possibly Northolt):
Despatch in pairs or wings, according to clouds, to patrol mid-Kent patrol lines at 20,000 to 25,000 feet to engage:

(i) Third or fourth wave attacks of bombers plus escort:
(ii) Retreating bomb raids of first and second waves:
(iii) To protect fighter aerodromes whilst the earlier Hurricane and Spitfire squadrons are refuelling.

Reinforcements from other Groups:
Immediately the enemy numbers appear to be more than 150, request two or three squadrons to cover the northern approaches to London, as directed in Controllers' Instructions No 7, dated August 27th, 1940.

These days of bad weather enabled our ground crews to carry out much needed servicing and repair work, but we were still short of aircraft and it was not until the 20th that I was able to get my hands on one to have a short, sharp 'refresher' flight using Hurricane L1572. It was a bit of a sluggard and I hoped I could get a better one. There was a lot of enemy activity this day but 501 was not actively involved. However, 605 and 603 Squadrons had several successes with a couple of 605's Hurricanes damaged, so our 'wing' had not been left out! The bad weather did not, however, restrict the enemy night raids with about 300 raiders giving London their worst blitz.

October 21st was another foggy day with poor visibility. I took up P5193 on a weather check but was recalled as Kenley clamped in and we were stood down. This inactivity was getting monotonous and we were fretting to be airborne.

The 22nd, 23rd and 24th October were the quietest days so far, any activity being mainly reconnaissance; we were not used and the weather was mostly foggy with rain and drizzle and very poor visibility. In these conditions our elevated location on the Surrey Hills was a disadvantage, as the mists and low cloud 'hung' over the hills and restricted our flying when other airfields lower down were able to operate, albeit with difficulty.

During this period of little activity and quiet living the Station Administrative Officer, a theatrical agent in peace time, laid on some special nights for us in the mess – one entertainment by Ralph Lynn and the second a cabaret and sing-song by some of the Windmill Girls – who were quickly escorted back to London after a very few drinks! But we did learn some new songs, one in particular about 'Hitler's Secret Weapon' ... and wondered if indeed he was aware of it?? This light relief was welcome as, even when not operating intensively, we found relaxation difficult, but it did give us time to catch up on events.

By now the impotency of the night defences was obviously becoming a major problem and the commercial and industrial affairs of the country were being greatly disrupted. This was emphasised to some of us when one night we made a sortie to London's club-land and we had to dive for shelter in the Piccadilly Underground Station during a heavy raid. It was packed with shelterers, men, women, children, some with their pets. On makeshift beds, blankets, pillows, with their essential belongings and food beside them. Playing, sleeping, eating, talking, love-making ... all the trappings of a home life in closely packed masses of humanity. Hot and stuffy, the smell was – to us fresh air fiends – frankly terrible and the air posivitely germ-ridden. The ventilation was curtailed and the entrances and exits reduced with watchful wardens covering them, but the spirit and morale were obviously unbroken.

Our reception was unbelievable. Though some asked, 'Why the – we were not up there after the b – ', others offered room to sit, coffee, tea or a beer, 'duckie'. You name it, they offered it!! This tightly gathered mass of humanity sheltering there was something the world should know about, we felt, especially the Yanks in their smug, cosseted, safe isolation ... Their spirit and determination laced with that wonderful cockney humour with a trace of spice in it.

With the 'All Clear' given, we made our way towards the exits.

One chap accidentally knocked into an elderly lady. She looked up with half-closed eyes and gave him a dressing down in lurid terms that included his ancestry. Then, opening her eyes fully and seeing him properly, she said, 'Gor, love you ducks, I wouldn't have called your Mum and Dad to accounts if I'd seen you proper! God bless you and get them effing Jerries!'

CHAPTER EIGHT

The Final Phase

October 25th found us up to strength in aircraft with the squadron still commanded by Squadron Leader H.A.V. Hogan, DFC, with Flight Lieutenants E. Holden, DFC, and D.A.E. Jones as flight commanders and Flying Officers N.J.M. Barry, V.R. Snell and E.B. Rogers. Pilot Officers K.W. Mackenzie, DFC, P.R. Hairs, K.N.T. Lee, DFC, R.C. Dafforn, Witorzenc, Kozlowski, E.G. Parkin, and R.S. Don, who together with S. Skalski had returned after being wounded in July. Flight Sergeant P. Morfill and Sergeants J.H. Lacey, DFM, P.P.C. Farnes, R. Gent, A. Whitehouse, T.G. Pickering, V. Ekins, L.J. Patterson, A. Glowacki, just back from hospital, W.B. Holroyd, with Flight Lieutenant J.A. Gibson and Sergeants D.N.E. Mackay, Crabtree and Green non-effective from previous wounds. Pilot Officer Goth, a Czech, had just joined us.

With improved weather, but heavy overcast, the prospect for some trade looked good and the build-up was reported early when we came to readiness at 7.30 with large formations reported to be boiling up. We were scrambled first at around twelve o'clock with the CO leading, climbing to 30,000 feet over London and being vectored towards a large formation. Shortly after reaching 30,000 feet the CO had to pull out with oxygen trouble and Blue 1, Flight Lieutenant E. Holden, DFC, took over the lead. The enemy formation was sighted over the Thames Estuary, below us, indicated by bursts of anti-aircraft fire. They turned east and split up; we also split up and half the squadron turned right to intercept breaking into the attack.

I cut across too far to the east. It was more difficult to judge speeds and distances at 30,000 feet, and I overshot. Two of the

Bf109's got onto my tail and I saw them in my rear view mirror just before they opened fire and as someone called over the radio, 'Break, Mac, break'. I pulled round, greying out as I did so, nearly blacking out, and both of them shot past me. I pulled back on to them and got onto the tail of one of them at about 200 yards, giving him a long burst from behind.

There was nil immediate result but the pilot released his hood which narrowly missed my aircraft, and baled out. He seemed to be delaying opening his parachute, or was waiting until he was the right way up to do so. I had to break off watching him as two more became too attentive. Losing them with a juddering tight turn and pull-up, I was joined by Pilot Officer Snell and we both latched onto another of them at close range from slightly astern opening fire almost simultaneously and it blew up in the air, disintegrating completely.

A fantastic sight: one moment there was an Bf109 there and the next just a large cloud of smoke and some debris and then nothing, literally nothing. Snell had watched the first pilot bale out and thought he came down somewhere near Sheerness. With no more trade, and the squadron split up, we flew back together.

In the same interception Flight Lieutenant E. Holden dived vertically after a Bf109, but it escaped into the middle of the cloud layer; he shot up again but found nothing and dropped down to just above the cloud layer. He spotted eight Bf109's milling round in and out of cloud at about 7,000 feet. With a magnificent height advantage and attacking out of the sun he gave one or two short bursts at point blank range from astern; bits flew off it and the aircraft gave a convulsive shudder. He had to break off then as he was being attacked by two 109's and escaped into the cloud cover.

After a quick, late lunch, we came back to readiness, scrambled again at around 14.30 on patrol over Kent at 28,000 feet. Toni Glowacki was back on operations and was a 'weaver' ... or 'tail end Charlie' on this occasion and I was leading a section. In good weather above cloud we were vectored onto a large formation of Bf109's at different levels, giving mutual support to one another, mixed fighters and fighter-bombers.

Crossing over a small formation with a good height advantage, the CO told me to attack with my section, in such a position it looked too good to be true! I called the section, 'bandits below at 2 o'clock, attack, attack, Go' and half-rolled to go down, as my port wing came up there was an almighty bang and another aircraft took my port wing off. My No 2, the new Czech pilot officer, who so far had always been lagging behind anyway, had clobbered me fair and square. The impact was like hitting a stone wall; the wing had torn off at the root, the aircraft flicked over and over, juddering and shaking as the engine raced at full bore and I could not throttle back, the throttle was jammed.

I was being thrown about like a pea in a bottle; in the cockpit the noise was deafening and the vibration exhausting as the aircraft tried to revolve 'round the propeller'. In a few minutes the engine tore itself out of its mountings and I had absolutely no control over the remains of the aircraft; I jettisoned the hood – luckily it was not jammed by the impact – and it came away cleanly.

The lighter remains now wallowed slowly like a leaf as I released my harness and waited until I was upside down to drop out, taking in large mouth, or rather lungs full, of oxygen as we were still very high. Pulling off my helmet, as it was still connected to the oxygen tube, I fell out and watched the remains fluttering above and away from me. Indeed it seemed to be falling more slowly than I was.

The silence was uncanny after the shaking and thundering noise following the impact and damage. I decided to do a delayed drop being so high and knew that I could open the parachute as I entered the cloud which was at about 16,000 feet and there would be ample height left; also at this time some of our pilots were being shot up by the enemy.

The freefall was indescribable, toppling over and over. As we had never been taught 'sky diving', now universally popular, I was conscious of the need to wait until the parachute was uppermost so that it would stream unrestricted and not come up between my legs – as another pilot had experienced and nearly broken his back by the sudden jarring reversal. Reaching the cloud top, I pulled the rip cord when facing the earth and was pulled up with a jerk that took the wind out of me, as I had been holding my breath. I was slowly lowered, or so it seemed, into a thick moist mist, getting quite wet. The descent was, or seemed to be, slow and the silence more uncanny than previously. I gripped the shoulder shrouds and eased

The author's Hurricane after forcing the Bf109 into the sea off Folkestone, October 1940

501 Squadron Pilots at readiness, RAF Kenley 1940. Sgt Holroyd, Sgt Farnes, the author, Sgt Lacey, Sgt Laws.

Pilots of 501 Squadron, RAF Kenley, October 1940. Author second from right, sitting.

501 Squadron personnel at RAF Kenley, November 1940. Author seventh from left seated.

(*Left*) Heinkel He 111 under fighter attack, Summer 1940. (*Right*) Dornier attempting to evade RAF fighter in summer of 1940.

Bf109E brought down on Friday, 25th October, 1940, in Sussex. Pilot was 'captured' by a farmer.

The author leaving Kenley for Filton, December 1940. Bristol Freighters on the left. 501 Squadron. Hurricane MK1.

501 Squadron pilots at readiness, Kenley 1940. *From left to right:* Holroyd Farnes Author Jones Lacey Laws Henn Skalski Konraad Dafforn Muchowski

myself up to settle more comfortably in the harness and came out below the cloud into warm air and sunshine from the west.

The countryside looked so green and peaceful as I drifted down, the wind drying me out, but the silence – just a swish of the air as it spilled out of the silk – was quite unbelievable ... until I noticed that I was drifting rapidly towards a line of high tension cables. By now I could hear the earth's noises, a distant sort of rumble, cars, people, animals. I tried to steer away from the cables, but only succeeded in making myself spin and girate and sway below the 'chute. I cleared the cables by about 200 feet, with a feeling of relief, and finally drifted down towards a fruit farm and landed in a small hay rick, a beautiful soft landing. Sliding off it, I was met by the inevitable – yes! – farmer with pitch fork!

Convincing him that I was friendly, he took me to his farm house for a cup of tea. I asked if I could use his 'phone; he said, yes, if I paid for the call! After a few words he agreed I could use it free! Big deal. So I contacted Kenley and requested transport, giving them the address and a map reference as I still had my map in my boot top. They confirmed, but told me to wait until they had picked up Flying Officer Snell and Sergeant Whitehouse who had also baled out. So we had lost four aircraft and Goth killed (apparently he went in from 28,000 feet) all for one Bf109 probably; not a good exchange. PO Goth's aircraft was subsequently recovered some thirty years later, buried in a ploughed field. We had nothing to celebrate that evening when we returned late after being picked up.

October 26th was again cloudy with bright periods, ideal for the raiders and they took advantage of it with south-eastern England kept on the alert all day, starting with attacks at high altitude at around ten o'clock. Snell, Whitehouse and I did not fly as the Doc wanted to check us over. This done, we were cleared for operations the next day. The squadron, however, had some trade and Ginger Lacey claimed a Bf109. Otherwise uneventful, though London was now being hammered nightly and seemed to be the main target for the bombers which had been found wanting in the daylight attacks.

That evening we heard a priceless story, told by our delightful 'Queen Bee', a tall Canadian flight officer. It seems that there was a dance at Caterham and most of 'her' girls, WAAFs, had gone to it; they were at this time billeted in several large detached houses near the airfield. Whilst doing her nightly rounds of inspection with her orderly corporal, they found one of the girls in bed in one of the dormitories, which were large old-fashioned bedrooms with equally large old-fashioned wardrobes.

When questioned why she was not at the dance with the others, the WAAF said that she came back early as she was not feeling well! At that moment there was a muffled cough from one of the wardrobes, or so it seemed, and the door of one of them opened suddenly and a naked body sprang out wearing a gas mask (we had to carry them everywhere in those days), clutching its clothes under its arms, the clothes clearly being those of an NCO, together with a pair of shoes.

Taken aback, she and her corporal – as she told it – stood aghast with surprise and a sense of wonderment, her immediate thoughts being, 'How the devil can I hold an identification parade, as we did not get his name?' The mind boggled. I doubt if much came of it, but the shoes gave a clue to the NCO probably being a pilot. The Queen Bee was reluctant to divulge anything more, except to comment upon the ingenuity of the Poles, and their quick thinking, and after all, they did have a way with the girls!

The next day, October 27th started fair but became cloudy as the day wore on. Raids started at 07.45, on London and the docks, with many sweeps in the afternoon in two main attacks; the weather favoured the enemy. This was to be the penultimate major daylight attack of the Battle of Britain, though we did not know it at the time. Our first scramble was at about 08.45, to patrol the East Grinstead – Maidstone line at full squadron strength with 'weavers' in conjunction with 'Turkey Squadron', with the CO leading and myself as his No 2.

Reaching 26,000 feet, we followed 'Turkey Squadron' as they were vectored onto an enemy raid coming in from the south-east. Turkey Squadron attacked first, into the middle of the formation; we followed and broke up to select individual targets. The CO dived very fast after a Bf109 as I went for another, but lost him as I avoided another Hurricane.

Pulling up hard from 20,000 feet I was attacked by a Bf109 from the starboard quarter: I pulled up and looped to try and get onto his tail as he shot past and found myself in cloud with sunlight to my left. I reefed round to get into the clear again and did so in a shallow diving turn at about 23,000 feet and levelled off.

There were several aircraft scattered round the cloud banks and I saw a yellow-nosed 109 emerge from a cloud below me at about 16,000 feet going west. He did not see me and I dived steeply and with a great speed advantage from above and behind him going below him directly astern to pull up and gave him two short bursts from 150 yards.

He developed a glycol leak immediately and then flames licked out below the pilot behind the engine. I throttled back as he appeared to be out of control going down in a gentle spiral. I followed him to 4,000 feet as there were more smoke and flames along the cockpit. He entered a thick haze and I broke off as he was obviously finished, the position being east of base near Tonbridge. Subsequently found and confirmed.

This time we rejoined and formated back to base with two missing, who returned separately. The CO clobbered his 109 but had to break off when his throttle jammed as he chased him in the clouds, streaming glycol. As several 109's were reported down in the Tonbridge area that day, his was probably one of them.

Two more scrambles were flown that day, on long patrols at altitude, neither being successful, the raids being well scattered and the cloud cover very much in the favour of the raiders, but frustrating for us and the fighter controllers. However, on the second scramble we had Sergeant Toni Glowacki weaving; he was seeing bandits everywhere and twice we broke unnecessarily, only to reform. Upon landing back at base without any trade we jumped on Toni – a very experienced Polish pilot with eyes like a hawk – and told him, 'For God's sake shut up until it is absolutely necessary!' 'OK! OK!' says Toni.

On our third and last scramble, we were jumped by a pack of 109's, without loss or trade due to the cloud, at about 20,000 feet without a word from our weavers. Maybe they had not seen them?

Reforming, we returned to Kenley after over an hour's patrol and again on landing jumped on the luckless Pole. 'Didn't you see them?' 'Why didn't you warn us?' etc. Toni, with a laugh: 'OK, OK, sure I see them, but you silly b – say shut up, so I shut up!'

Toni was commissioned later in the war and successfully

commanded a Polish squadron. After the war, it is said that he got his family out of Poland in a small compartment fashioned inside a load of timber on a railway long low loader. Still with the RAF he emigrated to New Zealand and joined their Air Force, later joining their Civil Aviation Department. Regretfully he died in the late '70s. He never flew without an evil-smelling bright red velvet scarf impregnated with the most exotic perfume imaginable – God knows where he got it, though we made several suggestions! You could smell it before he entered the crew room. But one hell of a fighter pilot!

That night we listened to and watched the night attacks and heard that our score for the day was a total of eleven enemy against two of ours; the balance was getting better! At least by day, but the task at night was yet to be resolved. Apparently 85 and 247 Squadrons had attacked two night raiders caught in the searchlights, without success. This was noteworthy as 247 Squadron still had biplane Gladiators at Roborough, near Plymouth. We were warned to expect to be called upon to operate at night if necessary and were ready to do so. In fact after my experience at night when with 43 Squadron at Usworth, I thought it might be interesting now with the excellent 11 Group radar and controlling.

October 29th brought us to readiness at 07.30 and we were first scrambled at 08.30 on a long patrol over Kent. No trade. Though the weather was good it was hazy with poor visibility as we kept to 28,000 feet. We returned to base after 1 hour 25 minutes; a quick refuel and some hot cocoa to warm us up and back to readiness. The second scramble came at 10.30 to cover the London approaches at 26,000 feet. Already a large force of raiders had been engaged and we saw many dog fights being fought out, but it was too late for us to join in and control kept us on patrol at altitude for another long flight, this time 1 hour 40 minutes and at the end of our endurance. It was like not being allowed to join in the game!

Back to base and another rapid refuel, some grub and back to readiness to be scrambled once more, this time to the south towards Tangmere at altitude, vectored onto another large dog fight, but again too late and no trade. The Bf's were now obviously prepared to stay and mix it, blow for blow, trying to eliminate our fighters. It was not working, however. This patrol was yet another long one before we returned to base after a frustrating, cold flight, whilst the Spitfire boys had most of the trade.

We returned once more and refuelled to be finally scrambled again at 4.45 p.m. to 23,000 feet over the coast. Well placed, we watched them being attacked by Spitfires and broke ourselves to attack in sections. I broke with Blue section and we split to take anything we could latch onto; I spotted a Bf109 attacking a Spitfire and opened fire at him from about 300 yards to put him off, from the rear port quarter and slightly above, closing fast from our height advantage. He half rolled away and dived to immediately pull up sharply into a steepish climb. With my high closing speed I followed him tightening the radius as we went up and, getting below and behind, gave him a long burst from 200 yards down to less than 100. He belched black and white smoke and fell away in an uncontrolled spin. I followed him down to 9,000 feet but had to break off with violent evasive action on being attacked from behind by another 109, only becoming aware of him when my rear view mirror disintegrated; he was being attacked in turn by my number two, Pilot Officer Machowski, who later confirmed seeing my first victim crash just inland from Dungeness.

We were out of the battle now and I marvelled once again at the way one could be in a milling mass of aircraft, fifty or so, dog fighting one minute and then an empty sky, as if there was no war at all.

We got back to base and called it a day, it had been one! We heard that 602 had had a field day and that the 11 Group tactics and controlling had been exceptionally good, 19 enemy being destroyed for the loss of seven with probably more enemy casualties on landing back in France. A good but tiring day, and we looked forward to a good night's rest.

October 30th, low cloud with almost continuous rain and drizzle and the morning quiet at first and we were at thirty minutes' call. Later in the afternoon I was on an airtest when the squadron was scrambled at about 15.30 to patrol the Maidstone area at 26,000 feet; I was not allowed to join them but to tag along and keep a look out. They got some trade at about 25,000 feet over Maidstone, which I watched from about 28,000 feet as they mixed it below me.

Turning back from the coast I passed over several friendly and enemy aircraft and noticed a lone Bf109 at about 22,000 feet making for the coast. I half rolled and dived after him with full boost, gyro sight 'on' and guns to fire, the usual routine. Closing very quickly to 150 yards, I gave him a long burst from the rear quarter down to 80 yards or so and saw strikes in and around the cockpit. He broke away and went into cloud and I lost him and returned to base. Later it was confirmed in the sea, but the pilot was probably rescued.

The squadron had had a reasonable amount of trade for one day, Ginger Lacey again getting one over the Maidstone area and damaging a second; cloud cover made definitive attacks difficult.

So October closed with enemy activity lessening by day after the last frantic fighter-bomber and fighter attacks, but the night raids increasing in ferocity and accuracy. The 31st being a stand down with little activity, we took a well-earned rest.

Thus ended the Battle of Britain proper; we had survived and the Luftwaffe had been severely blunted and Hitler's invasion cancelled for all time, his 'Sealion' abandoned. The daylight raids became more of a nuisance value whilst the night Blitz increased and spread to other cities, the night defences were found wanting still. Day fighter squadrons were now called upon to carry out 'fighter nights', a hazardous and unsuccessful task for a single-engined, single-seater such as the Spitfire or Hurricane without any interception aids or indeed training in the role.

October had been no less a strain than the previous heavy bomber daylight raids with fighter escort. In fact October had been a very telling period, due to the high altitude of the operations and the greatly increased speed and mobility of the large Bf109 fighter-bomber and fighter sweeps no longer tied to the relatively slow pace of the heavy bombers. The vigilance and operational activity had been greater and essential to survive or attack the enemy. The total volume of sorties and operational flying had been increased by the necessity to keep standing patrols over Kent and the continuous workload was telling on pilots, operations staffs and the controllers, without whom we could not have succeeded in intercepting raids. The ground crews and supporting staffs worked round the clock to give us the aircraft and weapons with untiring vigour and cheerfulness. It had been a wonderful team effort.

Long term, the balance of air power over Britain had been assured and the island base made secure for the future.

One subject of discussion now was, 'Thank God, Bader and his crazy Big Wing tactics policy, backed by his group commander, was soundly rejected by our group commander, Air Vice Marshal Keith Park, MC, DFC, who prevailed in the effective use of his fighters supported by the AOC-in-C Fighter Command, Sir Hugh Dowding. The slower reaction time and waste of effort forming up into Big Wings would never have met the tactical requirements to combat the hordes of fighter-bombers and fighters coming over in October, or the essential flexibility to tackle them as we did, albeit missing many for various other reasons.

We had time now to reflect on, and discuss, tactics, guns, sighting, training and always the efficacy of the .303 calibre machine guns came in for considerable criticism, the hitting power, or lack of it and the harmonisation. We all felt that the essential training in weaponry and air firing had been sadly neglected and that the attrition rate could have been greatly increased if pilots had given priority to getting into close range. Those of us who had been reasonably successful had always made a point of doing so to be effective; get it right first time; you may not get a second chance and it conserves ammunition. I was determined to go ahead with my ideas for a crude armament trainer.

It was also clear now that whilst so many pilots took part in the Battle and made up the squadron, these could be separated into 'killers', 'near killers' and the passengers; all contributed. An example was one of our flying officers who once returned from a scrap without firing his guns and, on interrogation by the Intelligence Officer, was explaining that he had not seen anything of the battle, when his rigger rushed in and interrupted with a frantic, 'Come and see your aircraft, sir.' Upon examination it was found that there were in fact 36 strikes on it, including four through the propeller! The pilot was unaware that he had been attacked. He was soon posted back to Flying Training Command, a lucky fellow.

So what now? The reduced tempo was difficult to accept at first, but we wondered what was coming next, which way would the war go, what form would it take, what was the future for the squadron?

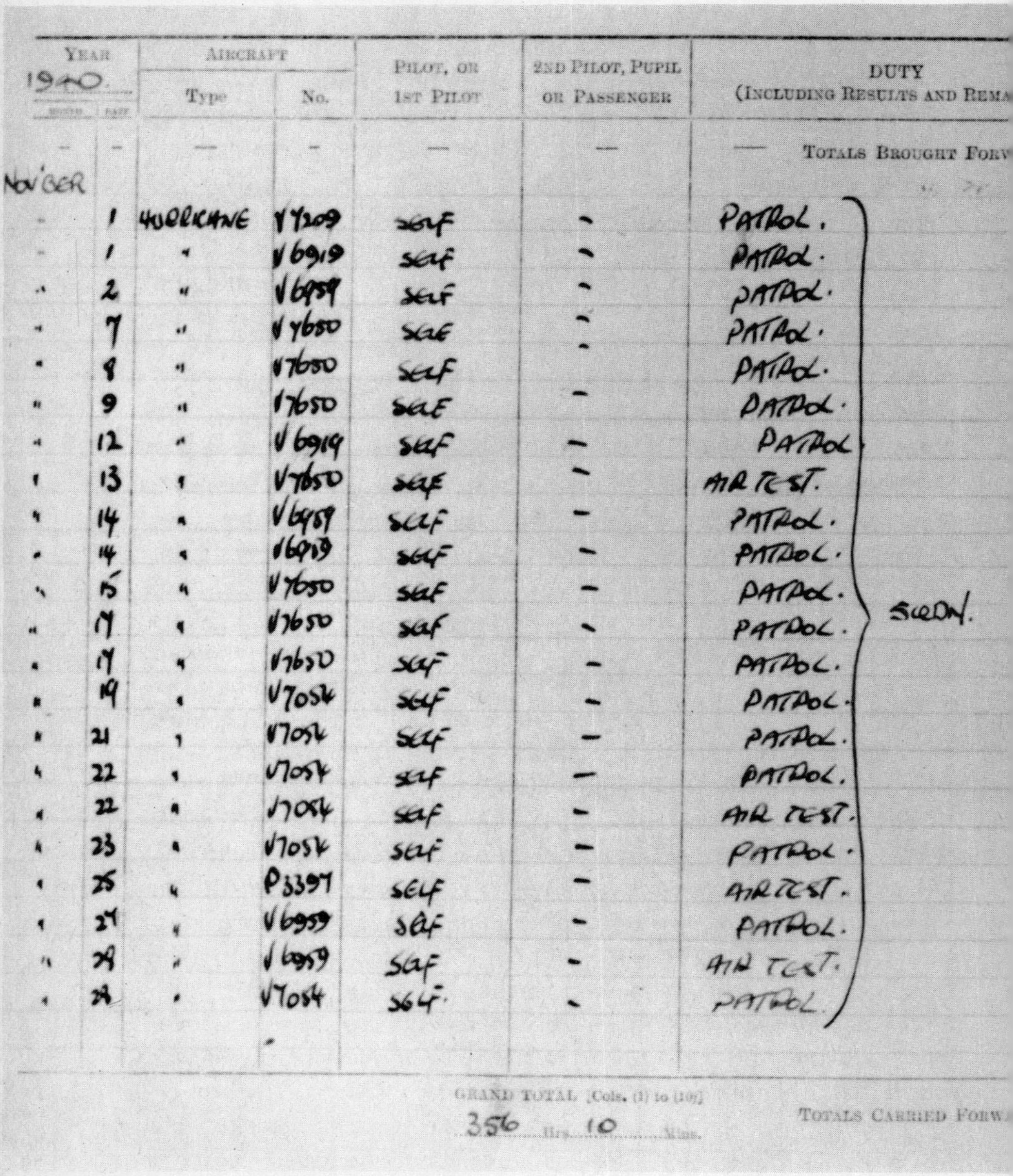

Year 1940 Month	Date	Aircraft Type	Aircraft No.	Pilot, or 1st Pilot	2nd Pilot, Pupil or Passenger	Duty (Including Results and Rema[rks])
—	—	—	—	—	—	— Totals Brought Forw[ard]
Nov'ber						
"	1	Hurricane	V7209	Self	–	Patrol.
"	1	"	V6919	Self	–	Patrol.
"	2	"	V6959	Self	–	Patrol.
"	7	"	V7650	Self	–	Patrol.
"	8	"	V7650	Self	–	Patrol.
"	9	"	V7650	Self	–	Patrol.
"	12	"	V6919	Self	–	Patrol.
"	13	"	V7650	Self	–	Air Test.
"	14	"	V6959	Self	–	Patrol.
"	14	"	V6919	Self	–	Patrol.
"	15	"	V7650	Self	–	Patrol.
"	17	"	V7650	Self	–	Patrol.
"	17	"	V7650	Self	–	Patrol.
"	19	"	V7054	Self	–	Patrol.
"	21	"	V7054	Self	–	Patrol.
"	22	"	V7054	Self	–	Patrol.
"	22	"	V7054	Self	–	Air Test.
"	23	"	V7054	Self	–	Patrol.
"	25	"	P3397	Self	–	Air Test.
"	27	"	V6959	Self	–	Patrol.
"	28	"	V6959	Self	–	Air Test.
"	28	"	V7054	Self	–	Patrol.

(Duty entries bracketed: Sqdn.)

Grand Total [Cols. (1) to (10)] 356 Hrs. 10 Mins.

Totals Carried Forwa[rd]

From the author's log-book for November 1940

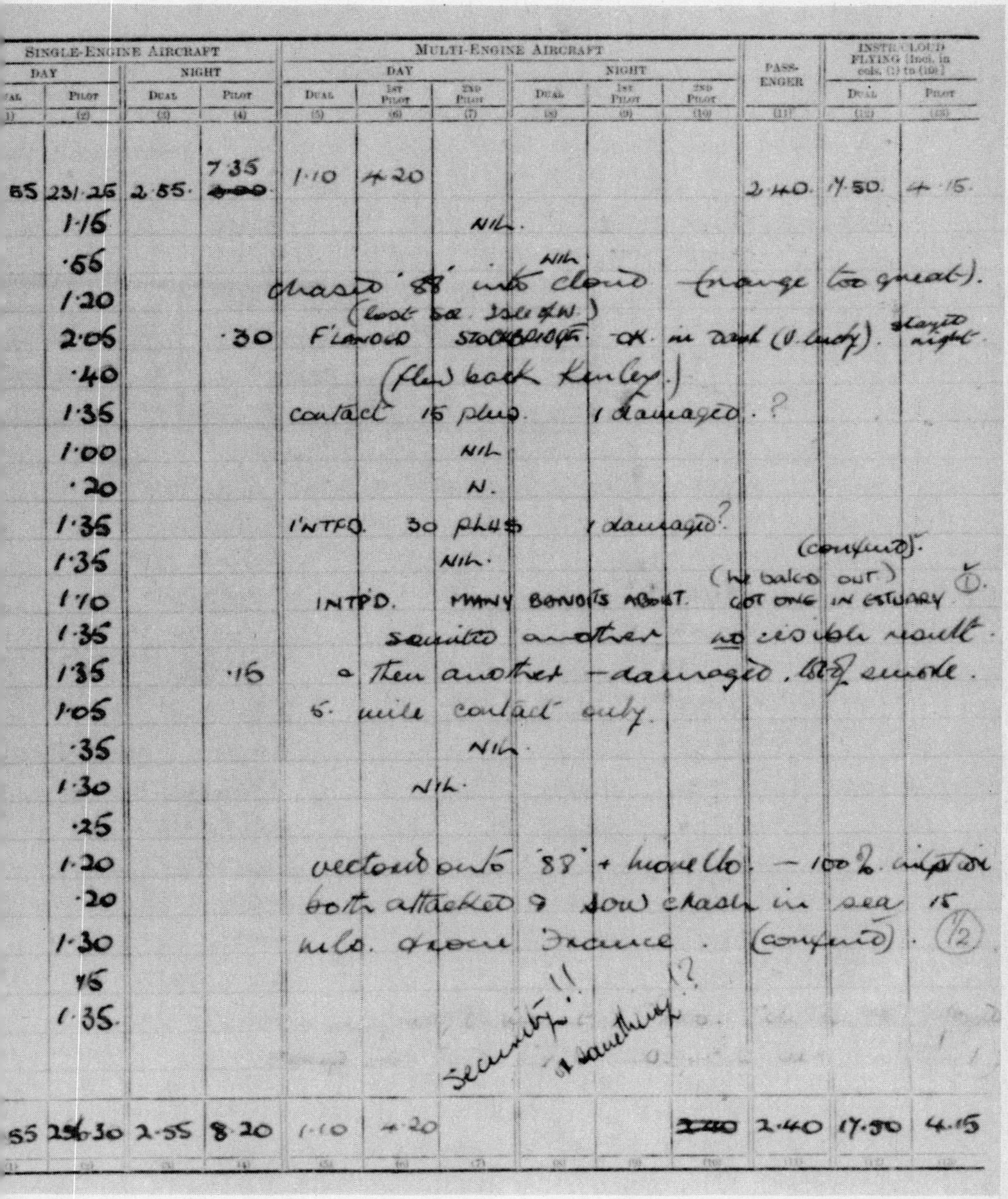

Single-Engine Aircraft Day Dual (1)	Day Pilot (2)	Night Dual (3)	Night Pilot (4)	Multi-Engine Aircraft Day Dual (5)	Day 1st Pilot (6)	Day 2nd Pilot (7)	Night Dual (8)	Night 1st Pilot (9)	Night 2nd Pilot (10)	Passenger (11)	Instr/Cloud Flying (Incl. in cols. (1) to (10)) Dual (12)	Pilot (13)	
55	231.25	2.55	~~8.00~~ 7.35	1.10	4.20					2.40	17.50	4.15	
	1.15												NIL.
	.55												NIL
	1.20												chased '88' into cloud (range too great). (lost sa. 3 sec of W.)
	2.05		.30										F'LANDED STOCKBRIDGE. OK. in dark (V. lucky). stayed night
	.40												(flew back Kenley.)
	1.35												contact 15 plus. 1 damaged. ?
	1.00												NIL
	.20												N.
	1.35												I'NTPD. 30 PLUS 1 damaged?
	1.35												NIL. (confirmed).
	1.10												INTPD. MANY BANDITS ABOUT. GOT ONE IN ESTUARY (he baled out) ①
	1.35												squirted another no visible result.
	1.35		.15										& then another – damaged, lots of smoke.
	1.05												5. mile contact only.
	.35												NIL.
	1.30												NIL.
	.25												
	1.20												vectored onto '88' + Morello. – 100% interception
	.20												both attacked & saw crash in sea 15
	1.30												mls. from France. (confirmed). ½
	.45												
	1.35												Security!! or something!?
55	256.30	2.55	8.20	1.10	4.20				~~2.40~~	2.40	17.50	4.15	

We had heard of new and better Hurricanes and Spitfires, heavier armament, improved performance and more modern equipment to combat the inevitable latest products of the German aircraft industry. At least we could now enjoy the undamaged facilities of a permament RAF station in close proximity to all the trappings of a large residential area.

It almost assumed the life of a peace-time station, with occasional exciting sorties providing the spice of life during the day, but the nightly heavy raids sharply reminded us of the seriousness of the whole business. To date I had never heard an expression of doubt about the outcome from anyone, service or civilian; all were unspokenly united in a common effort and bond. What a great pity this could never be invoked in peace-time.

November came in with the RAF consolidating and replenishing its forces, new and better aircraft, pilots with more initial training, the formation of twelve new squadrons. We wondered if we would be able to take the daylight battle to the enemy, get at him on his home ground with the attackers initiative. Not yet. The most significant improvement was the priority given to the night defences, the introduction of the night fighter Beaufighter with Airborne Interception Radar, not yet perfected but soon to be so. There was also better performance by the ground defence organisation, the anti-aircraft guns, balloons and searchlights, all getting into their stride. This improvement we could watch at night from our 'balcony seats' as it were at Kenley, overlooking London.

The saddest loss to us was possibly when our AOC-in-C., Sir Hugh Dowding, was replaced and virtually tossed aside after his magnificent contribution as head of Fighter Command before and during this crucial period. It was only now that all of us began to assimilate the facts, to really know what had been going on, what we had been contributing to, each in a small way.

The weather was rapidly deteriorating into winter. With shorter days, we were again warned about possibly being called upon to take part in 'fighter nights'. Already six Hurricane squadrons and two Defiant squadrons (withdrawn from day operations after being decimated by the Bf109's earlier) were designated night fighters and flying regular night sorties with single aircraft in co-operation with the ground defences and the usual fighter controllers. We looked forward to taking part if necessary and began thinking about ways and means, though few of the pilots had in fact flown the Hurricane at night.

The 1st November brought us two scrambles, the first one without any trade but the second had us vectored onto a raid near the Isle of Wight, probably going for the radar stations, given as a small formation of Ju88's at 6,000 feet. We found them in broken cloud, broke and attacked in pairs from above but were hampered by the cloud. They saw us and broke into the cloud layers; I attacked one with my number two and fired at long range. We lost him in cloud and after reforming and continuing our patrol we returned to base after 1 hour 15 minutes on station. The weather was favouring the raiders and visibility very poor and no trade reported.

November 2nd brought more bad weather and we came to readiness at a 'respectable' hour, nine o'clock, put back to thirty minutes as there was little build up. Later we were scrambled towards the Isle of Wight again at 10,000 feet in broken cloud for another long patrol. Towards the end we were vectored against a gaggle of Ju88's, sighting them fleetingly. We broke in pairs and searched after them in the cloud breaks. I chased one but lost him almost immediately; it was really difficult to see them in the murk. We gave it up and made our way back to base in ones and twos being unable to reform.

The next few days were quiet. We were stood down, 'weathered in' and devoted our time to domestic chores, socialising and personal matters. There were several personnel changes in the squadron, taking place and due. Squadron Leader H.A.V. Hogan DFC, was going to be promoted and be replaced by Flight Lieutenant E. Holden, DFC. Flight Lieutenant E.V. Morello had been posted in from Egypt, where he had been instructing at the SFTS, and Jones had been posted together with Snell. We missed the latter's piano sessions. Flight Lieutenant Morello brought a superb 1934 Meadows engined Frazer Nash and a black cocker spaniel, Lassie. We immediately teamed up with a mutual interest in sports cars and he took over 'B' Flight with myself as his deputy and I was told 'to break him in'! The ground crews had rectified most of the aircraft snags; we had received several replacement aircraft and were back to a full operational level. But where was the enemy?

On November 7th we were scrambled late to patrol the Southampton area in dreadful weather in which we were forced to split up, disperse and make out own way back to Kenley, a very long patrol with no trade; though the controller gave the enemy strength at 70-plus on Southampton, we saw nothing.

After two hours I forced landed near Stockbridge, being unable to find a way through the cloud now down on the deck, stayed the night at a satellite and returned the next day. Several pilots had been diverted, but none lost. London, Liverpool and Birmingham had been heavily bombed that night, with many casualties. A quiet day followed. I worked with the armourers on my trainer idea; they were becoming more enthusiastic ... a very conservative lot, armourers!

November 9th brought us some business, a scramble over Kent being vectored onto 30-plus together with the rest of the Biggin Wing. Weather was again broken cloud as we attacked in strength to break them up. With my No 2, I attacked a Bf109, as they broke into our attack, from astern and above and gave him a short burst at about 150 yards, strikes noted as he took evasive action and disappeared into cloud leaving a trail of black smoke. The dog fight was a desultory one for us all; several near misses as we dodged in and out of cloud in an inconclusive engagement with few successes. I claimed one Bf109, damaged.

November 10th and 11th were quiet days; Squadron Leader E. Holden, DFC, took the whole squadron up on a cloud penetration training exercise; I did not take part. On the 11th, I think it was, we were visited by the Father of the Royal Air Force, Lord (Boom) Trenchard and several of his staff officers. He came round the dispersal huts and we were introduced to him, all fascinated by his remarkable bushy eye-brows, his vigour and piercing eyes; a memorable occasion.

The whole daylight scene was now settling down and we were relaxed, taking stock, visiting London by day as well as night and enjoying the social life we had missed. 'Spike' Morello and I did some 'pairs flying', dog fighting and formation aerobatics and cloud flying. On November 12th we took off as a pair, with him leading, to patrol the Ashford-Maidstone line and look out for a lone Ju88 snooping around. The weather was good with some broken cloud at 8,000 feet. After a short search at 7,000 feet we were vectored towards him and spotted the Ju88 at about 8,000 feet, slightly above us to our left, exactly as informed by the controller. We pulled up to reach him, or get above him as 88's were their fastest bombers. We split up to 200 yards apart. He had seen us too and started to climb and turn towards the coast; we were between Ashford and Folkestone at the time. With full power we outclimbed him and he levelled off, jettisoned his bombs and when they had all tumbled out he went into a shallow dive towards the coast.

I 'stood off' as Red 1, Flight Lieutenant Morello, attacked on the inside of the turn and above, the Ju88 used his dive brakes and Red 1 overshot. I was then positioned for a beam attack from above, closing fast as his speed decayed due to using his dive brakes. I opened fire at 200 yards; no return fire was experienced and when I was down to 100 yards the starboard engine burst into fire and smoke. I attacked again from above at 100 yards as he went into wispy cloud; Red 1 joined me.

The French coast was dimly visible ahead as we followed him out of the cloud down to sea level with both engines now smoking and the starboard engine still on fire. We kept about 500 feet above him as he seemed to be levelling off very close to the sea as though to ditch. Suddenly he pulled up violently, almost vertically and stalled, sliding back tail first into the sea ... a remarkable sight.

This tail slide we discussed many times afterwards, wondering and conjecturing how it occurred; pilot or crew action, failure of controls, bouncing off the sea, or what? It was a fairly common aerobatic maneouvre at a suitable altitude and in the right type of aircraft, but a truly remarkable ending for such a relatively large aircraft as a Ju88.

We circled the wreckage, watching our tails at the same time as we were half way across the Channel, informed control and after some minutes returned to base. Spike Morello had been well blooded; that night we 'tied one on' well and truly! The following day, Sergeant Holroyd, who had been on an air test in the same area, said that about the same time (15.20) he had seen bombs drop and explode east of Ashford, but was not sure of the exact location.

He was at around 3,000 feet and had not seen any aircraft and had wondered where they had come from.

The Ju88 was conceived as a twin-engined 'dive-bomber', but became a very versatile bomber and reconnaissance aircraft and later a formidable night fighter fitted with interception radar. Its performance was high, fast and rugged; it was well built, armoured, and difficult to shoot down with .303 calibre bullets. Like most German bombers, the entire crew were accommodated in a compact forward nose cockpit area and difficult to hit from dead astern. One had to attack from above or below or in a beam attack at close range to be successful. Indeed we often thought that the Ju88, with such a good performance, was never fully exploited; certainly not initially.

In the latter part of November, I flew sixteen operational patrols with only a few contacts with the enemy in which we intercepted small gaggles of nuisance raiders, Bf109's. In these we had some trade, the first on November 14th, when we intercepted 15-plus approaching London at 20,000 feet; attacking them at 17,000 feet we broke them up and I claimed one damaged.

Again on the 15th, we intercepted over the Thames Estuary; flying high at 26,000 feet; in reasonably good weather with many bandits about, we broke up a widely spread formation. I attacked one Bf109 from slightly astern and below, from 150 yards; hitting him in a long burst the pilot jettisoned his hood and baled out as the aircraft broke up. He opened his 'chute soon after and was drifting down over the estuary as I attacked another, with no visible result.

Being attacked myself, I had to break off and dog fight my attacker, who was quite persistent; after several very tight turns I scraped round onto his tail. He took no other evasive action and really made it easy for me, firing into him at about 200 yards, well into his engine and forward of the cockpit. He slowed and belched smoke but I lost him as we went into a thick haze and I had to avoid a Hurricane. We did not join up after this fight and straggled back after a long patrol. This turned out to be the penultimate contact with our adversary, the last coming on November 28th. During this November-December phase, I claimed two and a half destroyed and four damaged, making a total to date of eight destroyed, four destroyed halved and four damaged.

December was quiet, with eight patrols recorded leading a flight or section, three squadron formation patrols and two cancelled scrambles, terminating in a squadron formation on Christmas Day. A fitting end to 1940. During this last period we had time to visit theatres, take part in more sport and carry out maintenance on our various cars. This included taking the body off the derelict MG 'L' type four seater and fitting it onto Spike Morello's Frazer Nash, but retaining the Frazer Nash radiator and bonnet.

Our outings were often disrupted by the still heavy and continuous raids on London and the fortitude of the Londoners, after a month or more of battering, never ceased to amaze us and the frustrating impotence felt during them persisted. We met pilots from other commands, discussed our various 'trades', tactics and aircraft and were all the more convinced that we preferred the role of the fighter pilot to any other. A short life and a merry one, with a lot of freedom of action and scope for the individual. Every man to his own poison, they say. This was mine.

Changes were in the air and we had been alerted to return to our home base at Filton, Bristol, the peace-time location of 501 (County of Gloucester) Auxiliary Squadron. The Auxiliary personnel of the squadron were, of course, tickled pink; this was going home for many of them, to their own houses and families, businesses that they had left when mobilised. A happy reunion was expected by all of them. We also heard that two of the previous flight commanders might be rejoining us. This did not please us, as they were senior to us but vastly less experienced and had very little operational knowledge or hours to their credit. However, wait and see? I did not know either personally, but many of the original pilots did, who had served in France and through the Battle of Britain.

From 17th December, we began preparing for the move but continued operating at full strength. On 27th December, we received several Bristol freighters to carry the immediate necessities and those kept behind for operations, the Bristol freighters being most appropriate. We took off at full squadron strength, formed up over Kenley for the last time, flew round and gave them a squadron farewell fly pass over the airfield and set course for Filton. There we arrived in good weather, after half an hour's flight, to give them a squadron break and stream landing as an opener. The freighters arrived some time later.

CHAPTER NINE

501 Squadron at Home

Filton was a different kettle of fish from Kenley; an old and very small station and officers' mess with rather cramped, but adequate accommodation. The airfield was essentially built for the Bristol Aeroplane Company and goes back to the First World War. With the one runway and the perimeters of the airfield hard against the factory buildings and works to the east and up the hill to the north, the airfield was locally defended by anti-aircraft gun posts round the perimeter augmented by a large balloon barrage. These balloons were disposed so that when they were 'up', and they went to 6,500 feet, we had funnels as it were at either end of the runway through which we had to fly when taking off or landing. This was of course on clear, or reasonably clear, days only. In days or nights of poor visibility the balloons were kept 'close hauled', that is tethered to their mobile winch vehicles and could be seen bobbing about in the wind just above the ground, like large silver baubles on a Christmas tree.

Our dispersals were, however, on the north perimeter of the airfield at right-angles to the middle of the east-west runway; a long way from either end of it. This created problems because it was a long and slow route to taxi along in order to take off into wind and engines would overheat, whilst the time factor to scramble was critical. To overcome these adverse factors we frequently took off from our dispersal straight towards the south *inside* the barrage when it was 'up', making a tight ground level turn to port or starboard into wind and out through the 'funnel' and then form up outside the barrage or above it, depending upon their height. A tricky manoeuvre, but satisfying and time-saving without

501 Squadron pilots with portraits, RAF Filton, Bristol, 1941. *From left to right:* P/O Mackenzie. Fl/Lt Holcen. P/O Dafforn. P/O Lacey.

501 Squadron Officers, Filton, April 1941, author second from right, sitting.

(*Right*) Ian Gleed (L) and flight commander C 'Roddy' Rayner (OC A Flight) 87 Sqdn

Sq/Ldr Ian Gleed, OC 87 Sqn, landing at Colerne on 6th February 1941, during Press Facility demonstration of comparison flights Hurricane vis-à-vis Curtiss Tomahawk. Hurricane, P2798, LK-A, in all black finish except for blue nose marking and two blue bars on rudder.

Hurri IICs of 87 Sqn in starboard echelon.

PZ865, a IIC, 'The Last of the Many', being flown by P.W.S. 'George' Bulman.

Author's Spitfire I, RAF, Colerne, April 1941, with 'red hand of Ulster' insignia on engine cowling.

overheating the engines. It did have its shortcomings, however, as was proven later.

After arrival, we had three days settling in whilst keeping a flight on standby each day. Bristol, we learnt, had received several heavy raids by night, notably on November 24th/25th, December 2nd/3rd, which caused severe damage in the city, again on December 6th/7th and more recently on December 13th/14th when Avonmouth was also attacked. We were welcomed back with enthusiasm as many believed that we could defend the city at night. Unfortunately we were in no position to do so.

With a couple of days off I was flown up to Kenley to collect the MG 'PA' which I brought back to Filton; a nice drive for a change, with the hood down despite the cold, very refreshing. Spike Morello went up the next day for his Frazer Nash, now MG-bodied and a bit of a hybrid to fox the locals. Indeed it did so when we became friendly with the many sports car enthusiasts in Bristol, to whom the rebodied Frazer Nash was a surprise packet, with its enhanced performance not associated with the 'L' type MG it aped!

In a few days sure enough Flight Lieutenants Cridland and Malfroy, ex-501, rejoined the squadron, 'Cam' Malfroy, an ex-Davis Cup Tennis player, was welcomed, not the least because as a good pianist he could replace V. Snell, our tame Fats Waller, who had been posted! A sprinkling of new pilots arrived and I was made training officer for 'B' Flight as well as deputy flight commander, which I welcomed, as it enabled me to get in more flying. Ginger Lacey had recently received a bar to his DFM and had been recommended for a commission, as had Tony Whitehouse and Vic Ekins. Hawkeye Lee had left us, together with Gibson, Henn, Crabtree, Green and Mackay. Cam Malfroy took over 'A' Flight.

The many original Auxiliary personnel who lived around Bristol were given a royal welcome by their families and friends and the famous Mauritania public house and restaurant in Park Street, with its several bars furnished from the *Mauritania* liner, became a general meeting place. Those who lived within reasonable distance of the airfield were allowed to live out and the serious war-time tempo and atmosphere of the previous months was soon forgotten. Not that we were lulled into any sense of false security or reduced efficiency, but the strain and intensity had gone; there was a relaxed, almost peace-time ambience. This was usually shattered every week by the snoop raiders and night attacks, either on Bristol itself or the adjacent cities. It was an artificial existence in some respects; the realities of the war had not yet fully penetrated the West Country as it had London and other places. Unfortunately, this peace was being shattered at night and soon Bristol was to suffer some devastating night bombings and show that the West of England could take it as well as the Londoners.

January came in without any trade for us, by now well rested and ready for anything. I had got myself a good aircraft, V7054, and besides training new pilots was working on the 'Gunnery Trainer' which the Armourers had become quite enthusiastic about. It was going to be a bit of a 'lash-up' but might help in the absence of anything better. We were being well entertained by the local people and new friendships developed, with some sporting events and facilities available, especially rugby whenever possible. We fraternised with the Bristol Aircraft people and were welcomed at the factory from time to time.

Operationally we carried out sector reconnaissances singly and in pairs to get to know the areas that we were defending. Bristol, the Bristol Channel and its many important ports on both sides: Gloucester and the many aircraft and component factories around it. Trained new pilots, many sorties and patrols flown but few in squadron strength as previously, all mainly against convoy snoopers and sneak raiders.

The radar, Observer Corps and early warning in general were not as good in the West Country as we had experienced in 11 Group. It was a broader front and the long coast line enabled raiders to get in close under the radar by approaching low over the sea. Interceptions were negligible; on several occasions with low cloud and rain we heard raiders near Bristol without any retaliatory action being initiated or taken. On one occasion we called the Guns Control to enquire why they had not been alerted as we had actually seen an He111 below cloud just off the western perimeter of the airfield and not a shot fired.

On another such occasion we heard one orbiting above the airfield and two of us leapt into two Hurricanes and took off without parachutes, helmets – or permission – standing on the rudder bar with our backs braced against the seats to raise ourselves high enough to see out ahead, as we did sometimes when taxiing. By the time we got airborne he had disappeared into the overcast and we had difficulty in landing. An abortive and probably foolish effort, but we were fed up with the apparent ineptitude of the airfield defences and the fact that we had not even been alerted of his presence in our sector. We were given a well earned lecture about taking the law into our own hands. Our Group Headquarters, Number 10 Group, commanded by Air Vice Marshal Sir C.J. Quintin Brand, did however take note, we heard, and investigated the incident, so maybe something was achieved after all?

Early in January some of us were invited up to Kenley to a New Year Ball and Tony Whitehouse, now a pilot officer, and I were allowed to take the magnificent old Hawker Hind, our squadron hack. We set off late in the afternoon on the appointed day to land at Kenley just before dusk; at least that was our intention! As we neared the Surrey Hills from the west we found them and the area, including Kenley, to the north completely shrouded in fog north of the Hogs Back and Kenley reported that visibility there was a few hundred yards only.

Unable to get down and too late to return to Filton ... also reluctant to do so ... we made a forced landing in a large farm field on the slopes of Hogs Back, beneath a layer of wispy cloud and as the light faded. Parking the Hind near a barn we met the farmer, explained our predicament and got his ready permission to leave it there; he said that he would keep an eye on it for us too. We phoned Kenley and they very kindly collected us in a little over an hour or so.

We had a glorious evening and the next morning – not too early – when the fog began to lift, we were taken back to the Hind with a couple of ground crew to handle the starting as the Hind had to be hand-started by winding it up, like a clockwork toy! We could probably have managed it between us, but the flight sergeant insisted. 'Just in case,' he said.

We were in for a bit of a shock though, as on arrival and approaching the farm we looked up and saw a line of high tension cables stretched out towards the field. Then we saw that in fact we had landed under them without knowing it, and that we would have to take off under them again to get out of the field. Ugh! How lucky can you get? Another 'life' gone, so how many left?

We got the Hind out from near the barn by pushing it and whilst the ground crew looked it over we paced out the field along a line giving us the longest take-off run, with a good thousand yards it was not going to be a problem; though it was rough in places; and long enough to take off and clear the hedge and trees to the west. We got her started first time, warmed the engine up and taxied tight up against the eastern hedge. I opened up on the brakes, got the tail up, and we accelerated rapidly keeping the tail up to reduce drag, bumped over the field into wind and at about 75 passed under the cables. Once lifted off I kept level to gain speed and at 100 knots zoomed up over the trees, ample speed. All was well and we circled the farm and ground crew giving them and the farmer a 'thumbs-up' and set course for Filton in good visibility with only a few cumulus clouds about.

Reporting the incident on arrival there were no recriminations, as there was no damage except maybe to our pride, but bad news that the faithful and much loved Hind, a 'proper aeroplane', was going to be replaced by a wooden horror, a Magister Mk 2, two-seater trainer, in a few days. A sad day.

So far we had not had any operational 'trade' but it soon came in the form of us being instructed to commence fighter nights operating from Charmy Down, some 30 miles away, a satellite of Colerne on another hill to the north of bath. Number 87 Squadron, Hurricane Night Fighters, was stationed there, commanded by Squadron Leader Ian (Widge) Gleed, DFC. We looked forward to this and flew up to Charmy Down one day to look the place over and find out from the 87 Squadron boys what it was all about and if there were any particular snags. From what we were told it was not going to be anything like the fun we had expected; in fact the whole business of single-engined, single-seater night fighters was obviously not for many of our newest pilots, not yet anyway.

Night flying commenced in earnest on January 3rd, in cold wintry conditions with snow and poor visibility. We would fly our aircraft up to Charmy Down after a day's training or otherwise in the late afternoon. This also served as the routine, and mandatory, 'night flying check'. Land there, park the aircraft and find our billets for the night in commandeered houses on the Colerne airfield perimeter. Then go to the dispersal huts where the accommodation was in direct contrast to that at Kenley. These huts were dirty, stuffy, from choking coke stoves and ill-equipped. We sat around in Irvin jackets, or flying suits chatting, playing cards or discussing the task with the more experienced single-seater night fighter 'boys'; Squadron Leader Gleed; Flight Lieutenant Smallwood; Roddy Rayner, Roland Beamont and others.

The outlook for any success seemed grim to us day fighters who were accustomed to more or less clear skies, gaggles of enemy aircraft to attack, followed by usually uneventful returns to base, or with some assistance when necessary, good land marks to fly by and a host of diversion airfields available if needed. This fighter night business was a contrast which could not be readily expressed, it had to be experienced; relieved a little by learning of 87 Squadron's problems – as they too had been thrown in at the deep end as it were – straight from day operations some months previously.

We went from dimly lit huts into black, bleak, cold darkness, getting our eyes 'night adapted' for night vision which took several minutes to acquire, despite wearing very dark glasses for half an hour or so beforehand, groping our way to an aircraft, cold and damp, with little illumination to taxi out and take off singly, alone for once, to endeavour to find an invisible patrol line. This line on a map should be indicated on the ground by groups of flares; these were inevitably either not lit or invisible due to cloud or haze. Patrol lines were near possible important targets, such as Bristol, Portsmouth, Southampton on the one side, with Cardiff and Swansea on the other side of the Bristol Channel. Harbours and aircraft factories defended by searchlights and anti-aircraft Batteries augmented by balloon barrages up to 6,500 feet in appropriate weather and cloud conditions.

We were supposed to operate in conjunction with these ground defence forces, controlled by the Sector Operations Centre at 10 Group headquarters, Rudloe Manor. This co-operation was intended to use the available forces to the best mutual advantage under the prevailing conditions. Our patrol lines were agreed previously on the edge of the searchlight and gun defended zones. If directed into this zone or sighting a raider and requiring to enter the zone to attack it we would call them on our radio, and indicate our intention by signalling on our 'downward identification light'. This was to tell them to stop firing and give us freedom of action within the zone without fear of being shot up by our own guns. It did not always work out that way, however, and the so-called co-operation left much to be desired.

Once on patrol in a cramped, unheated, cold cockpit, a thick windscreen and large gun sight blocking one's forward view, without any airborne radar, the distraction of ill-lit instruments and general discomfort, it was a matter for good old 'eye ball' vision, and the chances of spotting an enemy raider were minimal. In moonlight it was indeed possible, if the positioning was right, to 'lock on to' a silhouette and follow it, but any lack of concentration or any distraction and you lost it with little chance of picking it up again. It was at night that 'peripheral vision' became even more important than in daylight. It was found that concentrated 'looking' or peering, trying hard to see as it were, was ineffective, but that scanning relaxed was successful and otherwise obscure sightings could be noticed and then locked onto, movement across one's scan observed. It was not easy.

In cloudy conditions the only chance of seeing anything was against the glare of ground fires or bomb bursts or searchlights illuminating bombers – in short, an almost impossible task. Occasionally an accurate vector from the controller might put you onto a bandit in moonlight if favourably positioned initially; even then the range of vision was limited.

Anyway, we started and soon adapted to the training. We had to, as our first fully operational patrol was on January 16th, when Bristol took a hammering. We put up a maximum effort, orbiting the target area at 1,000 feet intervals in broken cloud. I was at 12,000 feet above the balloons, which were at 6,500 feet, as we watched the explosions and fires searching for a silhouette or a sighting in the searchlights or glare of the fires, but saw nothing. Once I thought I had just missed another Hurricane, seeing what looked like the glare from exhaust stubs. It was, I found, a more tension-creating type of operation than the action-packed daylight flights previously enjoyed; the groping in the dark, alert for anything, yet not quite sure what to expect, and flights of the same duration seemed to be very much longer.

Recovery and return to base were a new experience and always something of a hazard, except in moonlight when it was a doddle. In average conditions in winter at night, dead reckoning navigation was essential, with some help from one's operations centre or base if they had any facilities. The approach and landing were another problem without the usual daylight visual reference points; keeping to a safe height because of the surrounding hills, searching for the flashing beacon which would be doused if any bandits were about; culminating in setting up a reasonable approach and 'finals' to land alongside the line of 'Goose Neck flares' on the runway, which were seldom visible above 500 feet. The Chance Light, a strong beam that could be flashed along the runway as one rounded out to land, could be used to quickly illuminate the runway. Positioned to the left of the runway threshold so that you could keep it in view doing a 'left hand' circuit, it was illuminated only for long enough for you to be safely on the runway and then doused. It had one serious drawback however; its brilliance immediately destroyed your night adaptation or night vision so that taxiing in and parking were made difficult after landing. We preferred the Goose Necks only if visibility permitted.

Another problem that troubled us for a while were the longer periods of pure instrument flying without reference to any external land marks, horizon or fixed reference points. Interpreting the complete attitude of the aircraft in space and its performance from the many and varied instruments, provided us with the information necessary to do so. Whilst all of us were trained in basic instrument, or 'blind' flying, our experience was the usual day fighter pilot's short climbs and descents through cloud with clear patches, or short weather bound recovery to base, never prolonged flying with sole references to the instruments and no sight of the ground except for a brief time when taking off and landing. Instrument flying requires relaxed concentration and complete trust in the data shown on the instruments, to the exclusion of all or any physical, or indeed mental, sensations or feelings associated with being part of the aircraft or the effects of centrifugal force, in turns, or the gravitational pull as in a climb.

This is contrary to many pilots' natural reactions and the fighter pilot's need to 'fly by the seat of his pants', to be part of his aircraft. The result was that our early long night flying sorties imposed an additional strain on us that at first was not realised. Since the war, instrument flying has received the attention of experts and is now almost a science in itself.

This was the 'fighter night' scene. We coped reasonably well and gradually took to it; on moonlight nights we revelled in formation flying, aerobatics and navigation using the rivers, water and roads for reference, all most satisfying. Landings, however, were not the precision, smooth affairs that we had perfected. The poor visibility, glare and sparks from the throttled exhausts – we had not yet got the anti-glare shields – and the dull 'Goose Neck' flares relegated us to somewhat unprofessional landings, 'wheelers' or rather bumpy arrivals. No more 'cat peeing on glass' greasers. All very chastening to the ego!

Whilst operating from Charmy Down, we continued training and operating by day at Filton, but with no serious trade. Sergeant Grummet was killed when he flew into a balloon cable; we had begun to loathe the things, though they did keep the sneak raiders up above important targets. The last week was abortive with desperate weather and no trade by day or night, but on one reasonable day we indulged in some unauthorised flying under the Clifton suspension bridge ... it just had to be done! We were mildly chastised for it, of course.

February came in with more bad weather: snow, mist, rain and low cloud; some bandits about – one He111 flew up Park Street in Bristol at low level with the rear gunner machine-gunning civilians and disappearing into the mist and cloud. Our flying was very curtailed and we worked on cars; I was well on with the gunnery trainer and hoped to finish it in March. We were visited, on one of the better days, by the Duchess of Gloucester, our patron, and we had a very sumptuous lunch in the mess and a parade of the squadron and aircraft for inspection at readiness and in working kit. A proud occasion.

On the 23rd, I scrambled with Vic Ekins as my No 2 and a cloud base of 1,000 feet, after a very persistent He111. First we heard him in the broken cloud snooping about, then Sector Operations got us off and we scratched round Bristol and the airfield as best we could, in terrible visibility, and down to the Avonmouth docks. Once we thought we saw a swirling hole in the mist and cloud and a shadowy shape in it, followed it but got snarled up in the mist ourselves and returned to base, never having been above 1,000 feet!

Later in the afternoon I was called, through our Adjutant, to go down to the Bristol Police Headquarters, where they had the crew of an He111 which had crash-landed on the mud flats at Avonmouth after being shot down by a gunner on a moored ship using a Vickers 'GO' .303 machine-gun. With others, the gunner had been alerted when the alarm was sounded in the dock area – a wonderful effort by this intrepid gunner.

The RAF and Army interrogators found out that the He111 crew had seen an aircraft, a fighter, in the low cloud and had turned away over the docks to get to the Bristol Channel! So maybe we had contributed ... but it was the gunner's victory.

On February 28th, in the morning, we took a section over Cardiff at 24,000 feet but arrived too late for any trade; the birds had flown. This was followed by another scramble in the afternoon to maximum height after a high flying 'recce' plane, we got to 30,000 freezing feet but he was still above us. God we were never so cold, sitting on our hands again to keep them active, to stop them going solid and numb. It was, however, a gin clear day and we could see for hundreds of miles, perfect for 'recce' photography of course. Unable to reach him we returned to base to thaw out.

In March, I was transferred to 'A' Flight as training officer with several new pilots. This move I did not fancy as the flight commander, Cridland, and I did not get on and it would be difficult as his deputy. Nevertheless, we made a go of it. The weather was brightening up at times and trade developing and on the 20th, on a semi-training patrol, I was vectored with my No 2 onto a Ju88 near Gloucester in low broken cloud. We saw him once but lost him near Cheltenham. We heard later that he had unsuccessfully bombed the railway station from low altitude. Little damage and he got away. Undoubtedly the early warning system, radar and controlling in our sector were hampered by many factors and were in no way as good as the 11 Group set-up, through no fault of the controllers.

About this time we heard that we were going to be moved to Colerne airfield, near Bath, where we had been billeted during our fighter nights at Charmy Down. The Engineering Officer wanted to have a look at the accommodation and technical facilities, so we went up in the lately acquired Magister. Colerne was a peace-time built station, reasonably large airfield with a maintenance unit alongside it which had its own small airfield. The dispersals allocated to us were right alongside a road and large, well-appointed, with blast pens on either side. The set up looked very sanitary, we looked forward to it; on a hill top with clear approaches, no balloons and a large runway. What could be better?

Back at Filton, we continued training and I finished off the Gunnery Trainer that I had designed and built with the help of the armourers. This was simplicity itself with a pilot's seat, stick and rudder-bar coupled to a gyro sight on a gimbal so that the sight could be moved by the stick and rudder-bar, as though it was doing so as part of the aircraft being flown by the pilot. Using this we had several of the scale model aircraft, made for aircraft recognition training, of the more common German aircraft mounted on long poles which could be moved by the instructor at known scale distances, marked around the trainer, to simulate enemy aircraft. The pilot, though sitting still himself, could range and move the gyro sight as he would do if flying in his Hurricane or Spitfire. It was in fact quite effective and an eye-opener to most regarding their mis-calculation, or estimation, of range, which is all the difference between a kill and little or no damage.

This must have been one of the first attempts to provide such a static trainer; in later years, of course, sophisticated similar trainers were devised by experts and used to great effect, but as usual a bit late in the day.

About this time Ginger Lacey had been commissioned and also been to the Palace for the Bar to his DFM; Paul Farnes had received his DFM, and I had been invested with my DFC.

Towards the end of the second week in March, two of us were going up to Charmy Down for a fighter night. The aircraft were already there, so we decided to go up in the 'Matchbox Wonder', the Magister. Taking off inside the balloon barrage which was up I did our usual 'down the airfield and turn left, into wind and out through the funnel' act. As I pulled sharply round off the ground and came to 'take off left rudder' I found my left boot caught in the flap lever. With almost full left rudder on I was unable to lift the port wing clear of the ground, even with full right aileron, and banged the port wing tip lightly on the grass-covered ground, not the runway. Recovering in time and freeing my boot by breaking the trapped strap we flew out through the funnel and on to Charmy Down.

Upon arrival I told the ground crew about it and their 'Chiefie' (flight sergeant) decided to inspect the wing and main spar, cunning fellow – he knew Magisters, and found the main spar cracked at the wing root. Another 'life' gone?

March also brought several 'X' raids and we used these for training scrambles for new pilots, together with convoy patrols in the Bristol Channel entrance and up the Channel. Once we were vectored onto a Bf110 near Cheltenham going for Dowty's factory, but we missed him. Unfortunately March ended tragically when Sergeant Laws caught his propeller on our 'cack-handed take-off up the funnels' and crashed nearby after collecting a balloon cable on the way. Around the same time we had a new Hurricane delivered from the maintenance unit at Aston Down, just north of us. Another Ferry pilot came down in a Magister to collect the first delivery pilot and on a gin clear day when all the balloons were close hauled except one which was stuck up, flew into the cable and was killed. This balloon was on the road to the officers' mess, behind the mess, and it was jammed in the winch with the balloon at 5,000 feet; the only balloon in the whole of the West Country and he had to get it! Doing a tight circuit in a light aircraft to save time he had not seen it, hit the cable with one wing and spiralled down it onto the road to the mess and burnt out. We reckoned that those b – cables killed more of ours than theirs!

Early in April Vic Ekins was commissioned and Paul Farnes posted to a Fighter Operational Training Unit as an instructor, all of which was of course celebrated in Bristol.

April also saw our move to Colerne begin on the 4th. Being a very independent squadron we were happy with the accommodation which separated us from most of the station. Several of us were in fact billeted in some lovely old Bath stone houses on the steep hill leading from the main London-Bath road to Colerne. Our hosts looked after us well, but generally did not see much of any of us. I had got to know a well-known MG agent in Bristol, one Marshall, who had a garage at the bottom of Blackboy Hill, where he had a fabulous collection of MG 18/80's. His enthusiasm has not waned, I believe. Anyway, he had a delightful Humber 16 Doctor's Coupé for sale at a give-away price; I bought it as our Flight gharry. On posting later, I sold it to one of the sergeants for £37 10s and after the war when 'putting up the Squadron Standard' we met again. I asked about it. 'Oh, Ah. I sold it after the war for £400!' Lucky man.

Shortly before leaving Filton for good we heard an aircraft orbiting Filton one morning above a layer of drifting fog coming in from the Channel. At the same time Filton Tower had a 'Mayday' call from this aircraft and they rang us up to know if we could help, as there was an RAF Hampden bomber overhead short on fuel, and 'sort of lost' and could we help, or did we think that he could get in through the fog. The balloons, all of them, were close hauled at the time as it was clear above the fog.

Remembering the incident with the He111 and the effect of his passage through the mist and low cloud, I offered to go up and try and lead him down. He gave the top of the fog at about 600 feet and some of Henbury Hill to the south of Filton was visible, but he could not make out the airfield or approaches. I took off and shot up into the clear sky at 500 to 600 feet through the fog, no trouble; it was in fact not as thick horizontally as vertically, or so it seemed – probably the effect of the sunlight above it.

Joining him at about 1,000 feet at about 160 knots, we nattered on the radio and he confirmed that he could waffle along at about 120 knots with everything 'down' using lots of power. This was OK and I led him round on a left hand circuit using the few landmarks visible and Henbury Hill as a guide, first on a 'dummy' run from the east slightly down wind, but the wind was light and it was better to land away from the Bristol Aircraft factory buildings than towards them. On the second we came round nicely with Henbury Hill on our left and sure enough dropping into the fog my passage cleared a tunnel in it and he followed me down and landed as I did a 'roller' and took off straight ahead to come round and land behind him. Meeting him soon after, he and his crew thanked me and the tower and we discussed the clearing effect, especially as the fog was not nearly as bad as it looked from overhead.

I had a look at the Hampden, an aircraft I had only seen fleetingly in the air; a nasty cramped sort of tandem seating arrangement, lethal escape facilities and not much of a bomb load. Not for me, thank you, and no envy of the bomber boys; theirs was a thankless task. After attention by our ground crew and refuelling, some grub and clearing fog as it was burnt off by the sun, they took off for their base in Norfolk, leaving us a couple of crates of beer for consumption later.

Another incident about this time really astounded us. On a clear day a Beaufighter was taxied out for take-off, not an unusual sight then, and with the balloons close-hauled we watched it taxi to the eastern end of the runway and took no more notice as it disappeared round the corner out of sight. However, when it was just airborne someone sensed a change in the engine note and called us out to watch again. We did so as it turned left back towards the airfield at about 1,200 feet with the starboard propeller feathered flying on the port engine only. Mm ... Mm ... interesting we thought, could be a prang here!! A tricky bit of asymetric flying at the best of times, more especially in a Beaufighter in this configuration. In a curved approach with height in hand we suddenly heard the port engine lose power and someone called the tower as we piled into a couple of cars and started down towards the runway. As we did so the pilot cut the port engine completely, raised the undercarriage and belly-landed on the soft grass alongside the runway towards the Bristol works, coming to a halt about half way down the runway.

We drove down to it as the fire and rescue vehicles and ambulance arrived from the tower. A small, very small figure emerged from the aircraft in civilian flying overalls and helmet walked over to us; as it approached us it removed its flying helmet and a cascade of female hair fell over her shoulders. Surprise, surprise! We asked if she was OK? 'Yes, thank you, any of you chaps got a – cigarette?' It was one of the intrepid ATA pilots. Our first comments were, well done, but how the devil do you cope with the rudder pedals and forces, especially on asymetric conditions such as just experienced. Oh, that is easy I always carry my back cushion to pack me forward! She refused our hospitality and was taken by the Bristol Aeroplane chaps to their test pilots' accommodation; after all it was their aircraft! A magnificent effort. We returned to our dispersals very humble chaps.

At Filton we met their famous test pilot C. Uwins, a legendary figure in aviation and Dick Northway, another test pilot and sports car enthusiast. Whilst we were there he discovered a 1928 Red Label Bentley Vanden Plas short chassis 3 litre, metal bodied tourer for sale in Newbury. I took him there one day in the MG 'PA' to look at it. There is was, with only 17,000 miles on the speedometer, virtually not yet run-in, Battleship Grey, in 'as new condition'. He bought it there and then from the lady selling it, not asking too many questions as it might have been embarrassing for her – possibly it was a son's or husband's. With log book and receipt for, dare I say it, less than £100, he brought it back to Bristol, the 'PA' having a job to keep up with it; a beauty. As the paintwork was slightly faded he had it repainted by a Bristol coachbuilder; six coats, each carefully rubbed down, with a finish to concours standard. I have often wondered what happened to it, when Dick was unfortunately killed testing, after the war.

April continued with training from Colerne, some Army Co-operation flights and several night operations from Charmy Down, now just across the road, in better weather. On the 12th I was vectored onto a He111 in moonlight over Bristol at 12,000 feet or so, spotted him and turned to follow, but soon lost him going south at the same height. Another daylight patrol on the 14th put me at 28,000 feet with a section of four after a high-flying recce plane; we saw him fleetingly at about 4,000 feet above us, out of reach. We needed a couple of Spitfire Mk2's on standby for these fellows, lightly loaded to get to their height. Additionally we were just not getting the early warning necessary to position ourselves right for the many single aircraft venturing into the West Country.

April also saw the beginning of the next phase of the war, taking the offensive on the fighter front, beginning by escorting daylight bombing raids on targets in France. Our first experience came on the 17th when deployed in squadron strength to Westhampnett. From there we rendezvoused with a force of eighteen Blenheims to escort them to and from Cherbourg, where our task was to neutralise the German fighters on their Cherbourg Peninsula base. Flying at 16,000 feet we covered them as they approached the docks and then orbited the harbour and airfield as they went in from the south-west in a curving formation attack from 14,000 feet. It was suicide; the Germans put up a Box Barrage which the Blenheim crews courageously flew through holding a steady course on their bombing run. We had to jink from time to time as some flak burst near us – no way to get shot down! In some ways the Blenheims' attack was pathetic at that altitude knowing the intensity and accuracy of the German flak defences. They made their run, there was absolutely no fighter defence and we escorted about twelve back having watched, helplessly, as the others were shot down. At least we felt that the tide had turned; we were taking the fight to them now.

The rest of April was uneventful, several squadron strength exercises and new pilots shaping up well. Cam Malfroy was posted to Group Headquarters and promoted to squadron leader. 'Daff' Dafforn got 'B' Flight; Pilot Officer Waine was killed in a stupid beat-up and Pilot Officer Lockyer was killed in a car accident when just going operational – a waste.

Bath turned out to be a beautiful city still with a rural peace-time feeling to it, except of course for the enormous Naval presence there. The weather was good and again life almost tranquil with the emphasis on training.

At this time we heard that we were going to re-equip with some Spitfire Mk 1's, and possibly some Mk 2's from Exeter, secondhand really, but no matter, they were to be collected as soon as possible.

It was, however, with some mixed feelings that I contemplated losing the faithful Hurricanes. The Mk 1, with the constant-speed propeller, was a wonderful fighting aircraft with an exceptional combination of rugged dependability, stability and ease of control. A stable gun platform with absolutely viceless flying characteristics, a forgiving aeroplane. In the hands of a capable pilot it was more than a match for the Bf109. Its only shortcoming, which applied to the Spitfire too, was the inadequacy of the .303 calibre machine guns even though there were eight of them. We, certainly I, would be sorry to see them go, they had served us well and saved many lives.

With time to spare, I took a Hurricane down to Exeter on 24th April to look the Spitfires over and learn the cockpit – just like a Hurricane in most respects, with a few controls different and in different positions, rather more cramped, less forward view on the ground, smaller, but one seemed to be 'more part of it' in consequence. After returning to Colerne and some lunch, Sergeant Thomas (Mark 2 – we had several of them!) took me down in the Magister, which had been repaired and returned to us (in fact it had not been as serious as first thought), and I brought up the first Spitfire.

Lovely to fly, tricky to taxi, lighter in every respect, better acceleration and climb, certainly not such a roomy cockpit and some of the ancillary controls not well positioned, but promising. The main snag seemed to be the abnormally heavy aileron control above around 250 knots; most of them still had the fabric ailerons unfortunately. They were in fact a fair b – ; faster than the Hurricanes they were good for aerobatics, stable, forgiving, capable of being side-slipped and precision-landed but the narrow undercarriage called for more care than the Hurricane. Forward visibility made fast taxiing difficult and they over-heated quicker on the ground having smaller area radiators. Flick maneouvres were out, whereas the Hurricane was good for flick rolls down to 400 feet. Anyway I liked them and the Colerne Wing Commandar Flying, Wing Commander C. Beamish, one of the famous Beamish family from Coleraine, Northern Ireland, took Ginger Lacey, Marsden and me down, in a Blenheim, to collect three more the following day. These we ferried back smartly and found them superb for close formation and a quick 'tail chase' exercise.

The remainder were soon collected and we commenced working them up but still keeping some Hurricanes for readiness and night flying. With these we were scrambled five times on the 2nd and 3rd, two convoy patrols and three aborted, no vectors and desperate weather. On one patrol, when operating from Warmwell on the coast, with Pilot Officer Brockhurst as my No 2 in low cloud, fog and rain, it took us twenty minutes at 200 feet or so to find base and land!

Back at Colerne, the highlight of the second week was when Ginger Lacey and Wing Commander Beamish took off in two

247 Squadron celebrating author's first night victory at Predannack. Author and the CO Squadron Leader O'Brian, DFC, 13th September 1941.

(*Left*) 247 Squadron celebrating f rst night victory, RAF Predannack. Flight Lieutenant Carver and author, 13th September 1941. (*Right*) Pilot Officer Price, 247 Squadron. He was killed on 31st August 1941.

D. G. Smallwood – 'Splinters' – who eventually rose to ACM Sir Denis. CBE, DSO, DFC. Seen here as OC, 87 Sqn in a Hurricane II.

Lagonda 1931 blown 2 litre at Mullion Cove. June 1941

'Ulster' RAF prisoners of war. Schubin, Poland, 1942 Flight Lieutenant McConnel, author, Squadron Leader Cooper, P/O Brown, Flight Lieutenant Thallen, Pilot Officer Ayton, DFM, Flight Lieutenant Buckley, Pilot Officer Hannah.

Repatriated prisoners of war in Gottenburg en route for UK, 1944 Squadron Leader Burns, Author, Squadron Leader Osborne – end of centre table

(*Above*) Spitfire MKXVIs Gunnery Squadron 203 AFS RAF Chivenor, 1947

(*Left*) Author with Spitfire XV RAF Chivenor, when OC the Gunnery Squadron, 1947

Spitfires to do some mutual dog fighting using cine cameras to assess the results. On the first sortie the Wing Commander could not shake Ginger off his tail, so they landed and changed aircraft. This time Ginger could not shake off the Wing Commander. Obviously one aircraft was vastly superior to the other. They called it a day and prepared to land back at base. However, whilst airborne on the second sortie the wing swung round 180 degrees from west to east. Ginger, a cunning so and so, checked the 'Wind T' in front of Flying Control, noted the change, and came round to land into wind. Meanwhile, the Wing Commander had joined the circuit to land in the same direction as he had taken off, not checking the 'Wind T' and landing down wind!

We watched, fascinated; this was going to be even better than the Beaufighter affair at Filton! Everyone turned out to see the outcome. Sure enough, they had not seen each other and met head on, landing in opposite directions in the middle of the runway. The impact was actually between two wings and not directly head on, ending 'arm in arm' with two wings locked together, with two very surprised pilots standing by the embracing Spitfitres, a love match ... well hardly! Ginger was, in his own words, Fireproof! It was May 15th and the new AOC-in-C Fighter Command, Sir Sholto Douglas, was due to visit us in the next day or so. The two damaged Spitfires were therefore parked, not 'exactly hidden', on the maintenance units' airfield adjoining ours, hopefully out of sight?

When he arrived the AOC-in-C insisted on driving round the MU's airfield en route to our dispersals; he noticed them but apparently said nothing to the staff officers accompanying him. Had we got off with it?

When meeting us at dispersal and being introduced by our CO he questioned Ginger Lacey about not getting many Huns lately. Ginger said, 'Well, not many about, Sir.' The C-in-C said, 'No, but what about getting Spitfires? ... Like those two on the MU airfield?' Ginger began, 'Well, Sir, it was like this ... ' but the C-in-C interrupted and, turning to the very embarrassed Wing Commander Beamish said, 'I know what happened, don't do it again, we cannot afford it!'

Later in May I was told that I was probably going to be posted to a Hurricane night fighter squadron as training officer initially and a flight commander later. Meantime we went fully operational on the Spitfires, lost our Hurricanes in dribs and drabs and continued training with infrequent scrambles against hostiles.

On the 19th we scrambled – in a bit of a flap due to the paucity of the early warning – to be vectored onto a Ju88 in cloud; they had got his altitude all wrong and we overshot well below him and lost him. On the same day Pilot Officer Barton was killed flying into another of those damnable balloon cables, this time near Weston-super-Mare, returning from a convoy patrol. Squadron Leader E. (Gus) Holden, DFC, was posted overseas and replaced by Squadron Leader Boyd, DFC and Bar, and I was posted to No 247 Night Fighter Squadron at Predannack, on Hurricane Mark 2's, together with Flight Lieutenant D.G. Smallwood of 87 Squadron with whom we had operated at Charmy Down recently. Ginger Lacey got 'A' Flight when Cridland was posted.

With a few days' leave in hand, I visited London in the MG 'PA' and looked round the secondhand car showrooms. Finding a lovely 1931 Lagonda 2-Litre Tourer at Bartletts in Pembridge Villas, we negotiated an exchange with a little cash adjustment. It was a roomy fabric-bodied sports tourer, four-seater, running unblown but with the 'blower' thrown in for good measure, to fit maybe when petrol was more abundant? In beautiful condition, I drove it back to Colerne at a steady 60 mph on empty roads; it went well thereafter on the 'Green Stuff'!

Thus my tour with 501 Squadron ended: a memorable time with so many fine people, pilots, ground crew and supporting staff. Our many and various, sometimes mad, episodes are still talked about at the annual Battle of Britain Fighter Pilots Association Reunions at Bentley Priory, Stanmore. At these we seem to have one of the largest remaining, or attending, contingents; survivors all.

The squadron record in France and the Battle of Britain is history. In later years the squadron was disbanded and the standard laid up in the Bristol Cathedral, but not before they had subsequently distinguished themselves in the latter years of the war. I could not have served in a finer squadron. They had fought continuously from their initial deployment to France in May 1940 until after the Battle of Britain in December 1940. Often battered, but never defeated, they had withstood the onslaught without a break as few squadrons had done and one of their pilots, Ginger Lacey, had topped the list of enemy aircraft destroyed. I was proud to have contributed.

What of the so-called Battle of Britain? First and foremost, no one should be under the delusion that many of us really had the faintest idea of what we were up against, or what was really going on. Far from it. We were in the main a happy lot, a band of warriors if you like, whether officers or sergeants, with a job to do; enjoying the flying and the challenge without a thought or doubt as to the future or the politics of it all. We wanted to 'get stuck in', get an enemy or two and back to base in one piece, with a serviceable aircraft.

We lost pilots, yes, and aircraft, but usually replacements turned up. On many, many, days it was hectic, nerve-racking, physically and mentally exhausting, but at the same time exhilarating and stimulating. No heroics and but little, if any, sentiment ... a job to be done to the best of our ability.

Since then there has been great emphasis and play upon fear, stress, emotion and 'the pressures'. Of course, all these were present, but it was all part of the task we had undertaken of our own free will and why not? Pressures and other emotions are all part of life and should be accepted as such. The subsequent exaggeration given to these elements by writers, and the Press, who were never near the participants, has been disproportionate to an embarrassing degree; a product of the mass media which over-reacts and overplays every possible human endeavour or action. Sensationalism where there is none, maybe courage and bravery and a lot of professionalism by all concerned in the air and on the ground, but let us get it in the right perspective.

What of the Allies who joined us? Frankly they were relatively few in number, except possibly the Polish contribution, and they mainly arrived fairly late on the squadrons in the Battle anyway. This does not belittle their magnificent contribution in either numbers or experience, but again their contribution has, in my opinion, been exaggerated. The Poles had been blooded before us, but the language problem mitigated against our gaining very much from their experience.

Regarding the aircraft, so much has been written about the

Hurricane and Spitfire that anything further would be superfluous. I flew both machines and liked them both, but preferred the Hurricane, as mentioned previously, for its ruggedness, stability and relatively light controls with viceless flying characteristics. It could take anything that you could give it and out turn anything flying at the time; well flown, a match for anything.

The Spitfire was smaller, lighter and slightly superior in all-round performance, but suffered from incompatibly stiff aileron controls above 250 knots. Both suffered from having normal carburrettored engines versus the German patrol injection system, much to our initial disadvantage when diving after them.

The Messerschmitt Bf109E, with which we were competing was, I consider, overrated but better armed, even after they seemed to abandon the cannon firing through the centre of the propeller boss. Again, so much depends upon the pilot and it seemed that it took more experience to get the best out of the Bf109 than out of the Hurricane or Spitfire. The 109 controls were undoubtedly heavy about 260 knots and their elevator control was vicious in fast steep dives and they had a tendency to flick near a high speed stall.

Our armament was, I consider, inadequate in hitting power and range, but the higher rate of fire with the standard harmonisation probably flattered a mediocre shot in registering some hits, though not lethal enough for a kill.

Our training proved satisfactory in all respects, except gunnery, but in many instances determination was lacking and the knowledge of just how much the aeroplane could take, or the pilot give it, when flown to the limit.

What of the pilots and squadrons? Some of both were in the forefront at all times. Squadrons because they were well led and had an élite who scored the most and lasted the longest; these lasted the full course of the Battle. Others took a hammering and fell out, to be replaced, these usually had a few 'killers' but insufficient to keep up the tempo. 501 was, luckily, one of the few with a generous share of killers and ended with one of the highest scores as a squadron. Many lacked the 'killer' instinct which is essential for success in any sport, or war, where physical and mental skills are combined to survive.

Finally, we were 'playing' on our home ground which gave us

some advantage; all contributed with a love of flying, unselfishness and a spirit that made one proud to have been part of it.

Of the tactics and results, these too have been analysed ad nauseam. Generally we were too busy and uninformed to be able to comment, but what little we did hear about the backbiting between sector commanders and, in certain instances, the blatant self-interest, sickened us. The subsequent treatment of our Air Officer Commander-in-Chief, whose actions in our opinion were fully justified and vindicated, was an insult and a disgrace to the Service.

The radar and controlling were as good as they could have been and saved the day of course, after which the mainstay was the ground crews and factory workers who kept us flying by producing the aircraft and armament; and the observers and plotters who augmented the radar. All contributed. The administrative services, equipment stores, medical, all did their jobs without heroics; we never really got down to analysing, or even discussing the so-called Battle of Britain. It was over, we had enjoyed it and were still alive, what else mattered, and whilst the war lasted there was another job, another posting, let's get on with it. At first the question of promotion, command, be it ever so insignificant, seemed to be of secondary importance to us volunteers, who by this time made up sixty percent of the aircrews. If it came we took it; if not, well one might not live to enjoy it anyway. However, we did begin to become aware that our apparent rapid progress, or success, was resented and many held back in consequence to make way for regulars who had the seniority but not the ability sometimes. Maybe the lull after the Battle allowed too much socialising and time to think, meet others and hear their problems; we had not had time for this sort of thing previously. It was a passing phase, we hoped.

CHAPTER TEN

Night Fighting Proper

After a spot of leave on posting, I drove down to Cornwall in the recently acquired Lagonda and found 247 Squadron settling into Predannack on the Lizard, the most southerly airfield in England – in fact the very *last* going south! Constructed upon marshy land on the moorland overlooking Mullion and Kynance Coves, the newly built runway was concrete and proud of the ground, so woe betide anyone running off the runway into the marshy surrounds. Accommodation was in Nissen huts, wooden offices and a small hangar or two – sparse in the extreme but adequate under the circumstances.

Personnel were, in the main, billeted locally in Mullion, Helston and Land's End. Full use of the Mullion Cove Hotel was made for the officers and most of the pilots were there, or nearby. The general atmosphere when I arrived in the middle of June was almost holiday-like, with visitors swimming, fishing and enjoying themselves, joined by the squadron personnel whenever possible. The food was superb and plentiful, with a surfeit of fresh fish and enormous lobsters, brought in daily by the fishermen.

One notable visitor from time to time was the famous A.P. Herbert and his family; he kept us entertained in the bar and at tea time with stories and anecdotes galore. Geraldo, the band leader, also stayed there frequently and gave us some music from the London shows; both of them characters in their own right. The fishermen we got to know well, but to them we were always 'furriners'!

No 247 Squadron was unique in that they were the very last squadron in England to be equipped with, and use, Gloster Gladiator bi-plane single-seater fighters. So equipped they were initially stationed at Roborough, the small airfield mentioned previously, defending the Naval bases at Plymouth.

Commanded by Squadron Leader P.G. St G. O'Brian, a quiet-spoken Canadian in the RAF, who had recently taken over the squadron, they had been re-equipped with Hurricane Mk 2 'C's and 'B's. The flight commanders were Flight Lieutenants D.G. (Splinters) Smallwood, ex-87 Squadron at Charmy Down and Johnny Carver.

Possessed of their new aircraft they were working up with several inexperienced pilots under training, their night fighter role being a new one and not yet fully developed. We came under No 10 Group at Rudloe Manor, still with Sir Quintin Brand as our AOC-in-C. He was a bit of a tartar about night flying and insisted that 'if the Hun is flying ... you can!' but sometimes forgetting that perhaps the Huns' airfields in France were serviceable when ours in England were not! It was said that he had incredible night vision, like a cat. Indeed he was once an unexpected visitor at Charmy Down in his Hurricane in very bad weather when we were not flying and gave us hell for not at least doing some training!

As there were five radar stations in the South Devon and Cornwall peninsula sector, we could expect more accurate and efficient controlling and appropriately better chances for interceptions and looked forward to some more trade.

The aircraft were painted matt black for camouflage in the night role and fitted with exhaust shields over the stub exhausts. These were essential to partially eliminate the glare of the hot exhaust stubs and reduce the blinding effect of back pressure with violent flame and sparks when throttling back on landing.

The long-range Hurricane Mk 2 'B's were fitted with additional underwing tanks and were armed with 12 x .303 calibre machine guns. The Mk 2 'C's had normal tankage but carried some of the first 20mm cannon, two in each wing making a lethal battery. However, they were still an unknown quantity and had yet to be 'fired in' and experimented with as they were known to be subject to severe stoppages. Eventually they proved effective and lethal with greater hitting power and range. Indeed, my most fervent wish later was that we had had them in the Battle; against the bombers the attrition rate would have been greatly increased.

Operations from Predannack were usually over the UK during night attacks on harbours and cities in our sector, convoy patrols round the coast and into the Bristol Channel. When there was no local activity, or as ordered, we carried out night and dusk intruder flights into France, code-named *Mandolins*, in moonlight when it was possible to carry out reconnaissance or attack selected targets and airfields.

Training was also carried out by day and night for the new pilots, together with some daylight operations from time to time. Once again the contrast experienced whilst doing fighter nights at Charmy Down became evident and the living standards, sitting round in the dispersal huts at night, trying to say the least. In other respects it was a relaxed time and the flying and operations not as intense as previously. We thoroughly enjoyed every minute of it, keeping fit with swimming or tennis in the afternoons when stood down, until the evening.

During my initial flying with 247 in June, I was averaging two sorties a day, or night and day, but absolutely no trade. On moonlight nights in reasonable weather the flying was superb, mutual pairs 'eyeballing', that is spotting one another and following by sight. Aerobatics and some formation flying. The manifest problems related at Charmy Down persisted and were perhaps more aggravating because they were now part of one's daily, or nightly, life, not just a diversion as when with 501 at Filton and Colerne. We accepted them, but every night sortie in a single-seater fighter in doubtful weather with few, if any, navigational or approach aids available is a bit of a hazard and a challenge.

I was made training officer and a deputy flight commander. With the occasional squadron 'Balbo' the morale was good; the life encouraged it. The patrols over cities and convoys were somewhat wearing in cold cockpits, concentrating on the instruments and there was a feeling of futility without the constant 'trade' I had become used to. The intruder operations (*Mandolins*) were usually pre-planned in moonlight or at dusk, rather more exciting and certainly interesting. They required accurate course keeping across the Channel, there and back, making a good landfall on the French coast for final navigation to the target or round the designated area being recced. When operating in pairs or more, the formation during the flight over gave no problems; these came after when each individual had to make his way back, as we seldom rejoined. Other such operations were singly in the Long Range Mk 2 'B's which also gave no problems, except for the inevitable illusion of a rough running engine as soon as we crossed the sea – soon dispelled by a quick look at the appropriate instruments!

July took on a more serious look with much more intensive training, patrols, and 'firing in' the cannon. As expected, these were giving trouble with their frequent stoppages, the breech blocks sticking and having to be honed down, a tedious task in which some of us assisted the armourers.

On the night of the 7th I was scrambled to cover a convoy in the entrance to the Bristol Channel at 8,000 feet. Arriving on station I overshot, or missed the convoy which was reported approaching Lundy Island. Orbiting Barnstaple Bay, I was told by Sector to watch for a bandit approaching from the north at an estimated 7,000 feet. I climbed to 9,000 feet to have some height in hand and circled down moon, but saw nothing and the plot faded according to Control as they also told me that there were several plots in the area and to look for another coming from the north, same height as before.

Turning to look for this one I spotted something against the low stratus cloud, about $\frac{1}{2}$ a mile away; he was turning south-east and travelling quite fast. I informed Sector and they confirmed a plot. Keeping him in sight I 'gated' after him, not losing sight of him now clearly silhouetted; I was catching him slowly, he must have been doing about 260 knots, in a shallow dive towards Falmouth. It seemed a long chase and the concentration to keep him in sight quite difficult, resisting the temptation to look at the instruments and flying by feel, but with a distinct horizon to refer to as I kept the bandit in sight.

Closing up on him, I was sure it was a Ju88; his speed supported this, and as we neared the coast Sector confirmed having us plotted. There was no cloud now and I wondered if the Falmouth balloons were up and if he might hit them.

Getting to within about 800 yards of the raider, he turned south-east, crossing the coast south of Falmouth at about 1,500 feet. Clearly identified now as a Ju88 I caught him up and gave him a burst at about 250 yards from below and astern and lost sight of him momentarily, but picked him up at about 100 yards' range. This time I gave him a longer, raking burst from below and astern; there was no result or indication of hits still and I throttled back to position for a third attack, knowing that by now I had little ammunition left.

As I did so, he started to burn and burst into a sheet of flame and broke up. I frantically took avoiding action, pulling up and away to the left. As I did so my windscreen was splattered with oil and some debris hit my propeller. Informing Sector I returned to base, landing in good weather, jubilant, very lucky, first at night! Next day we celebrated with a lunchtime session, in glorious sunshine.

After our celebration and a rest, we came to readiness again at six o'clock, with a clear night forecast raiders expected. Scrambled at 9.30 p.m. to patrol the Plymouth base line, the inland line, I watched a rather sporadic raid on the city and harbour area with considerable searchlight and anti-aircraft activity.

After about half an hour, with the moon well up, but no visual sightings so far, I spotted an He111 quite clearly illuminated, and then another, crossing from west to east. I turned in, called the guns and signalled my intention on my downward identification light, as I crossed into the AA zone the searchlight lost the bandit and the firing ceased.

I locked onto the second raider which was about 1,500 yards away, but had difficulty in seeing it clearly. Height was around 8,000 feet and closing rapidly on him I settled down to dive and get him from astern and below. When at 800 yards or so and slightly below, I had a very clear silhouette and, closing to 400 yards, was just about to go in for the kill when there were two shattering explosions on either side of me.

My aircraft was tossed about like a toy; I lost sight of the He111 and control of the aircraft and went into a dive, being momentarily stunned and disorientated. Immediately regaining control and getting back to level flight, I cursed the gunners who obviously had opened fire when they should not have and then suddenly remembered the Plymouth balloon barrage. Were they up or down? I was by then down to 5,000 feet, over the harbour. I pulled up into a steep climb and got back to 8,000 feet as quickly as possible, while calling Control and reporting the loss of a pretty certain kill owing to the blasted gunners. I was in a cold sweat, and furious.

The aircraft felt OK and I went back onto the patrol line but by then the raid was over, raiders gone, and the stupid gunners had nearly got me instead. More than just frustrating – inefficient gunnery control. We had a right royal verbal battle the next morning, and they finally apologised – not much help – and promised to tighten up their co-ordination instructions. Another life?

The rest of the month was quiet; training continued and 'firing in' the cannon, with some improvement in their performance. The slewing effect when two stopped on the same side before the adjacent two was quite surprising and, of course, ruined one's aim. The glare at night from the muzzle blast was dazzling and also affected night vision. The waiting periods in the dispersal huts when on readiness at night were wearing; some accepted them quietly, whilst others of us, myself included, were impatient for some trade. Monopoly was the chief occupation ... we must have become property millionaires each in turn, several times over. Nice thought, anyway.

We visited Helston, Falmouth, the Helford River and fished with the local fishermen whenever possible. People were still holidaying down from London and elsewhere and at times the war seemed very distant. In some ways it was; we were not in a target area. There was no 'station' life as usually associated with an airfield or RAF base and we felt somewhat cut off from reality. For the first time since being mobilised I joined in and supported the local vicar at his little country church on Sundays, together with many of the squadron, and we gradually became part of the Mullion Cove scene, accepted by the locals and visitors alike.

The 31st August brought tragedy. Johnny Carver and Pilot Officer Price, an ex-Army chap who had transferred to the RAF, found and attacked a Dornier 17Z snooping round the coast at sea level in the morning, with good visibility. Surprising the snooper, they attacked from above and behind, Johnny giving Price 'first squirt' as it was his initial interception, and following close behind him. They got the Dornier but Price apparently misjudged his attack, or froze onto the gunsight, narrowly missed hitting the Dornier and he flew into the sea near it as Johnny finished it off.

More intensive cannon-testing was carried out in the weeks following, with training and several scrambles. On the 5th, when on an air test, my IFF set (Identification, Friend or Foe) blew up and filled the cockpit with smoke; it felt as though I had been jumped by a fighter and shot up! Opening the hood to clear it away and everything else feeling OK with no visible fire, I hastened back to base as fast as possible, landing followed by the fire and rescue tender down the runway, but all was well, except for a burnt-out IFF set.

This was the first time that I fully realised the danger of fire in the air, of not really knowing what was wrong and therefore not being able to do much, if anything, about it. These thoughts made me realise how fortunate my flying career had been so far, several close shaves but never a serious technical failure as such. How many lives does one have?

August 8th, Pilot Officer Roberts was killed when making a forced landing following an engine failure and trying to get back to base just did not make it; another sad loss. When on pairs patrol with 'Doc' Doherty as my No 2 on the evening of the 25th, returning from a convoy patrol we were diverted to Portreath, our parent station, in bad weather. When approaching Portreath, we passed a couple of He111's going the opposite way over St Ives with light anti-aircraft fire from Portreath following them; they dropped some bombs near St Ives without any damage and we landed to spend the night there. We returned the next day.

It was a feature of this part of Cornwall that whenever Portreath, on the Bristol Channel above St Ives, was 'out' because of fog and sea mist Predannack, on the Atlantic, usually remained clear, and vice versa. This was fortunate for us, as long as one had sufficient fuel to divert at the end of the sortie. One night, returning to Predannack short of fuel, I found it in low fog but had sufficient fuel to risk diverting. With a lot of juggling, in almost nil forward visibility, but with extra Goose-neck flares and using the Chance Light for longer than usual as a guide, I scraped in, but ran out of fuel taxiing back up the runway. The luck of the Irish? Another life? With twenty-eight sorties recorded, August was a reasonably busy month personally and better for us all now that the squadron was fully operational. The only outstanding matter was the use of the cannons and the necessity to get the Hurricane Mk 2 'C's into the battle.

During these months 87 Squadron, with whom we had operated at Charmy Down, were instructed to deploy a flight to the Scilly Isles to fox the Hun and get at the convoy snoopers and possibly the four-engined Condors on their Atlantic flights.

These raiders had enjoyed a relatively free run of the area and sea approaches hitherto, penetrating low over the sea undetected. Our efforts to find them in bad weather had been unrewarding and trade poor in this respect. 87 Squadron had already initiated night intruding in Hurricanes operating from Warmwell, following their experience from Charmy Down.

They were taking the war into France in their own way, but they too found the short range and poor lethality of the .303 calibre machine guns virtually useless for ground-strafing. In May, they were active from the minute air strip on the cliffs above the town on St Mary's Island, Scilly Isles, where their detachments operated with considerable success for several months. They got some unusual German aircraft, such as Dornier floatplanes, a Dornier Flying Boat and others. Johnny Carver and I visited them one morning, but circling their tiny airstrip thought better of it and returned to Predannack! Before deploying there, they had practised on a 450 yard strip marked out at Charmy Down and perfected their short landing technique. 'Rubber' Thoroughgood, one of their pilots, told me when I met him that they had watched us and they were awaiting our efforts to land with interest ... and no little trepidation! Their detachment was code-named Operation Fishing – most appropriate, at least we were no longer alone in Cornwall!

August 29th culminated in a long *Mandolin*, intruder at night, with two Hurricane Mk 2 'B's, the long range jobs, covering reconnaissance of Lannion and Morlaix airfields down towards Brest, a look at Brest and back to Rennes and home. With Sergeant Doherty as my Number 2 we flew at around 2,000 feet in open formation, visibility good, after taking off at 22.00 and making a landfall at Ile de St Batz and thence to Lannion where there was no activity. Several lights were seen, some vehicles on the road from

Lannion to Morlaix which we followed towards Brest. Nothing at Morlaix, but approaching Brest our presence was noted by some flak-firing towards us; though we were quite a distance from them, the bursts were very high and they obviously thought we were bombers.

Skirting the town, which was clearly visible, we turned inland flying east towards Rennes; several beacons visible but little activity of any kind seen. As we were approaching Rennes at 1,800 feet, the lights of the town were quite bright and could have been mistaken for runway lighting, or a dummy flare path. From there, we turned west again and made our way to the French coast; passing near Lannion we landed back at Predannack at 01.30 after 3.30 hours of an uneventful but pleasant flight, but no trade.

September came in with shortening days, but reasonable weather. I took over command of 'B' Flight and Johnny Carver was promoted and posted to 10 Group. We were sorry to lose him. Training continued for the replacement pilots, but unfortunately we lost another, Sergeant Birkin, who went into the sea off Land's End during a practice 'dog fight'.

We persevered with the cannons on my Hurricane Mk 2 'C', carrying out four firing tests before being convinced that they were now working one hundred percent; this was BD832, a new aircraft and a good one. With the tests OK I looked forward to trying them in earnest. This opportunity was to come sooner than I expected.

About this time we heard that 'Splinters' Smallwood was leaving us to be promoted and get a squadron. We were happy for him, but sorry to lose him and his delightful wife, Jean, who was as much a member of the squadron as we were.

September 13th – we were not superstitious, whether it was a Friday or not – brought some trade in the form of convoy patrols. My last, at ten o'clock or so, was being ordered to patrol a convoy, or some boats going to join up into a convoy when in open sea, passing down the Bristol Channel west of Portreath. Before I arrived on station this was changed for me to patrol Predannack at 4,000 feet, but with no further information. Once overhead Predannack, I was ordered to 6,500 feet and vectored on a course of 300 degrees to cover a convoy forming north of the Scilly Isles, or beyond them, where activity was reported. I increased speed in good visibility but some drifting convection cloud at 3,000 to 4,000 feet; passing Land's End on my right, I saw some anti-aircraft fire away to my left, with tracers being visible. I 'gated' at full power to investigate, reporting the fire to the controller.

When nearing the ships which were just visible, or discernible as vague shapes on the water, I saw an aircraft come out of the thin cloud at about 3,000 feet below and to port (left). With a height advantage, I winged over gently keeping the enemy in sight and at about 320 knots rapidly closing on him. It was an He111 shadowing the convoy; with a perfect attacking position I closed to 600 yards and opened fire with the four cannon – and missed. The retarding effect of the cannon blast dropped the nose and my shots went low.

He returned my fire immediately, but was high, and what looked like two large orange Very lights came towards me as I opened fire from above and dead astern. As I fired up the fuselage and across into the starboard engine, this time allowing for the retarding effect of the blast, the He111 swerved to the right as the starboard engine stopped in a cloud of black smoke. I throttled back ready for another attack, as the pilot righted the aircraft and gave another burst, this time into the port engine. We were very near the sea by this time and as I broke off from only about 200 yards, bits and pieces broke off the raider; some hit my aircraft and tore the fabric in the fuselage behind the cockpit. I could feel the impact, but no serious damage. The He111 then side-slipped towards the water, ditching port wing first in a cloud of spray and settled nose down.

I reported it to Sector Control and flew round the spot several times but could not see any survivors as it slid deeper into the sea. It was difficult to see clearly anyway, but easy to fly round with a good horizon to guide me. The cannons were terrific, and hits seen easily, even at night, but the harmonisation must take into account this retarding effect or we will have to make allowances for it.

I thought about this when flying back to Predannack, where a rousing welcome awaited me; I remembered afterwards that of course when 'firing in' the cannon we had not actually been firing at a target; therefore, the effect of the blast upon the attitude of the aircraft had not been noticed. I believe that some bits of an He111 were washed up several months later.

This was my last operational sortie with 247 Squadron before ditching off France in the episode with which I opened my story. Almost a year to the day of operations and how many 'lives' lost? I make it seven. A total of 290 operational flying hours, over 300 sorties, including training and leave with ten enemy aircraft destroyed; 4 'halved' destroyed and 4 damaged. Still with two 'lives' or so left, it had, frankly, been a good year and I had enjoyed every minute of it. This may sound dreadful to the reader, but it would be hypocritical to say otherwise. Flying was my life, so were the risks, the challenge, the individualism. Flying a wonderful aircraft to the limit, as one with it, the excitement and exhilaration of combat ... short and intense. The team work of the ground crews, team work with the controllers searching in space for the enemy. We had not yet experienced the full weight of war, its horror and catastrophe, ours had been a relatively clean war, a very serious sport if you like, with no holds barred, and we had survived. What was to come? What next?

CHAPTER ELEVEN

... So There I Was Captured

Floating in this very small dinghy, wet cold and a little stunned, with a sore head, my first thoughts were to fill in the gap between ditching and finding myself in my present predicament. We used to fill in time when at 'readiness', or on call, by discussing experiences and actions to be taken in emergencies, such as baling out, or ditching, or possibly fire. We went over the actions required, what to do, how and when; mentally going through the motions until, in the event, it would be automatic.

This is what I must have done, released the parachute in its pack, inflated the dinghy by the CO_2 bottle and climbed into it. It was a desolate feeling, and situation, a hundred and twenty miles from base, from England; a hundred and twenty miles of formidable sea crossing to Cornwall with a very hostile coast behind me. What was the best thing to do?

It was around 21.00 by this time with a calm sea, a light wind from the west and a friendly moon, some broken cloud but visibility of a mile or so with the French coast outlined in dark silhouette. A few dim lights flickering here and there, their blackout was never the best, but no accurate idea of my position beyond the fact that I must be somewhere in Lannion Bay, or thereabouts with some small islands dimly visible against the horizon to the south. My first reaction was to get as far from the coast as possible so I put on the inadequate paddles. They fitted onto one's hands like mittens and it was soon quite obvious that the only way I could get any effective movement was by turning round and paddling backwards; this produced a reasonable 'rowing' stroke, but of course I could not see where I was going. By putting the small 'sea anchor' out astern the drag helped to keep me straight, but also required more effort.

Whilst paddling steadily I looked back over the day with mounting anger. Firstly the rather hurried briefing for this intruder operation, its implementation and execution. We had been told that a flight of Whirlwinds of 263 Squadron, together with an escort of nine Spitfires of 313 (Czech) Squadron, were being deployed to Predannack to attack Lannion airfield in Brittany in the evening and that we, (247 Squadron) were to follow in with two flights of Hurricane Mk 2 'C's, each with four cannon. To the best of our knowledge, and from very recent night reconnaissance flights by ourselves, there were no squadron aircraft as such based at Lannion, or any other aircraft, except possibly a Bf109 which was observed tipped up on its nose. Lannion had, of course, been used extensively in the Battle of Britain.

We had queried the efficacy of this operation, stating the facts as we saw them, and emphasising that the element of surprise would be lost in the initial attack by the Whirlwinds. Therefore we had taken off in a bloody-minded atttitude; I was angry about it and determined to give Lannion a good 'going over', and had reaped the reward!

Clearing my mind of these things and resting a bit, it was obvious that I was making little, if any, progress. I cleaned the wound on my head with salt water and suddenly remembered the many fishing boats the French kept moored quite far from the coast near their villages in this area, because of the abnormally long way the tide went out.

I wondered if perhaps I could nobble one of these and sail back? Turning round, I commenced paddling towards the coast in a southerly direction, resting every twenty minutes or so as my arms became like lead. The water in the dinghy had by now warmed to body heat and insulated me against the cold bottom of the dinghy; it was deathly quiet except for the lapping of the water as I made about two or three knots at the most.

After several hours it seemed – as my watch had stopped – the coast loomed nearer and larger on the horizon; it must have been after midnight by then. About one or two o'clock I could make out dim shapes between myself and the coast, backed by very dim light here and there.

Paddling on and on, looking round every so often, I saw the

outline of several boats and the glint of moonlight on metal at times; they were quite far apart. Coming up to them at last, the first was much too large and heavily chained to its mooring; the second was smaller but still too big, they both had motors. The third was quite small, about eighteen feet long with mast and a brown lug sail, no motor and the rudder in place. I got aboard, now quite warmed up, though with wet legs and bottom, but my arms were tired from paddling. I tied the dinghy on astern, cut the two mooring ropes with my knife and without much difficulty hoisted the lug sail and steered out to sea in a light wind, letting the wind drift me clear before setting course. The dinghy was holding me back so I hoisted it aboard; it was only about five feet long and two and a half feet or so wide anyhow.

Progress was slow and with the westerly wind I decided to try and get to one of the several islands off the coast nearby, but it was obvious that I could never get far enough out to sea before dawn not to be seen. There was quite a large island visible to the south with what looked like a sort of tower sticking up on it. By now it was about three or four o'clock. Still making poor progress, I was slowly getting near the island when the sky to the east began to lighten. Keeping as close to the wind as I could, I made for the island which was by now maybe half a mile away. Reaching it as the dawn was coming up, I moored the boat to some rocks in a cove; it was a craggy island, and took the dinghy ashore to use for shelter, it being very cold. Tucking it between a sort of low wall and a bush, I got into it ... and must have fallen asleep.

I next woke, or was wakened, being prodded by the rifle of a Kriegsmarine, one of several who stood round me with guns at the ready. Seeing that I was a bit groggy, more from cold and stiffness than anything else I think, one of them helped me get up. They took the dinghy and escorted me across the island to the tower, which was in fact a lighthouse just above a small harbour with a jetty where their patrol boat was moored; it was not an 'E' Boat, and was about thirty-six feet long with a wheel house and a 20mm cannon mounted forward and a cabin below. They were surprised to find 'Ein englisher Flieger' as nothing was known about an aircraft ditching and apparently they had been summoned by the lighthouse people, who had seen the boat approaching the island and thought it was some Frenchman making for England, or something.

After some hot ersatz coffee I was taken aboard the patrol boat, plus the dinghy. I gathered that the lighthouse people had been told to fetch the boat to the harbour later. After half an hour or so, we moored at a jetty at a fishing village or port, probably Rampol. There I was kept in the Kriegsmarine Offices for some time, given some bread and wurst and some milk. Interrogated in a vague way by a civilian whom I took to be Gestapo, in very bad English, I gave my name, rank and number, or rather wrote them down for him. This was followed by a Frenchman asking about the boat, in schoolboy French; I rather thought that he was trying to tell me that I should have hidden in it and they would have helped me. I could not really make out his position, or where he stood with the Kriegsmarine; all most peculiar.

I understood that I had to remain with them until someone arrived from Lannion to pick me up and take me there, after all I was Air Force! I was the first British pilot they had seen I think, as different people – military and civilian – kept coming in and looking at me as I sat in their, what seemed like an orderly room.

I was finally taken to a Kriegsmarine officer's office; a lieutenant, I believe. His English was better; name, rank and number again, nothing more, except was I aware of the penalties for stealing a boat? Whilst I was with him a Luftwaffe staff car arrived and I was whisked off at pistol point, in the back of it, to Lannion by road.

At Lannion I was surprised to be taken to their mess by an elderly Luftwaffe colonel and given a slap up lunch, with a delicious steak. Whilst eating it with him, he did not eat, he asked if the French mess staff could 'look at me'. They came in, looked at me and wanted to shake hands; all very touching, but I was in no mood to fraternise. After lunch he took me to his office, took down my name, rank and number and promised to let the Red Cross people in Geneva know that I was a POW the same day. I believe he must have, as I subsequently found out that the squadron knew within a few days.

He wondered why we had attacked Lannion again, but gave nothing away regarding the results of our attack. Then he referred to a wall map of the *Ost* (Russian) front appropriately pinned and marked with their, so far, rapid progress and surprised me by commenting that 'no matter who wins the war they will have to fight Communism ever after' and that 'Britain and Germany should join forces now'! This was later put to us as propoganda several times, but for me '*Der Krieg ist fertig*' ... and it was as far as active flying was concerned.

That night, my first in captivity, in a cold, bare cell in the guard room at Lannion was not a pleasant one. A hard wooden bed without a mattress, one chair, a toilet bucket and one blanket and a straw-filled pillow. Concrete walls, about ten feet by six feet, with a small window at the top.

Unable to sleep, I paced up and down, but had no room, so lay on the bed and contemplated how I had got there and what was going to be the outcome. The initial reaction was to chastise myself for being such a BF and out of the war just when it was getting aggressive; I had been given a flight, lost a perfectly good aircraft and why had it ended thus?

In flying, motor racing, anything involving ourselves with machinery and the elements there are many factors involved; training, skill, expertise in using the medium and one's skill to the best advantage, taking risks ... calculated risks which to the uninitiated seem foolhardy but which in fact are not. Seizing opportunities when they present themselves, quick reactions, calculating the odds; in short, making the most of what you have got in every way possible, but – and it is a very large but – there is an element of luck, call it what you will. It is a combination of conditions and circumstances that allow you to make the most of them; an indefinable something that some seem to have. I felt then, and later, that somewhere, somehow I had a very generous share of it. The luck of the Irish?

So, what had happened, quite simply I had allowed emotion, anger, to distract me, to cloud clear thinking and action and this was the result. In short, my own stupid fault. Never attack a ground target a second time round, and keep your cool.

Regarding being captured, this brought home to me the fact that those of us engaged upon intruder operations, at this initial stage of Fighter Command taking the single-seater fighter into action against the enemy on his home ground, were totally unprepared, untrained or even briefed on what to expect if forced to land in enemy territory. We were ignorant of evasion which was essential if one was to try to get back; once caught the chances were nil to slight. We were not equipped with any evasion or survival kit, or information. We flew as we did over England, in the briefest clothing, unarmed, and in fact the thought of not returning had never entered our heads! A sorry state indeed, someone, somewhere had not done their homework – neither had we as pilots – we were dropped in at the deep end, in more ways than one in my case! With hindsight then and later, I could probably have evaded and perhaps got back, who knows, or been helped by the French as the Frenchman had hinted. For the moment, it was an example of 'for you the war is over' Or was it?

CHAPTER TWELVE

Into Captivity

Entrained at Morlaix Station in a second-class carriage, accompanied by my middle-aged guard, we set off via Paris for Germany. The train was packed and more like a holiday train, with a large number of military personnel on it than a wartime train in France, except for the military presence at all the stations. Packed and noisy we made a swift run to Paris via Le Mans, where my guard bought me some peaches and grapes from a station vendor. The presence of a British POW went completely unnoticed.

Arriving at the terminal in Paris the chaos, the crowds, were like a Bank holiday rush in England. In the midst of this, my elderly guard got off in front of me. I allowed several people to get in front of me and seeing him disappear to the left up the platform in a mad stampede of passengers I went the other way, back down the platform towards the way we had come in, against the tide of passengers getting off. Quite unsure of what I was going to do, I ducked into a sort of railway workers' store with picks, shovels and a wheelbarrow with tools in it. Hiding behind the wheelbarrow, but leaving the door ajar, I waited impatiently, cramped and uncomfortable.

No one came near the store and after about half an hour or so I crept out, there were a few people about, and walked further down the platform and reached a side entrance with some steps down into a sort of yard. Walking down these, I was just moving off towards the entrance when a platoon of Luftwaffe appeared, coming in at the double. I turned and ran for the steps and heard a '*Halt*' behind me. I looked round and down to see the Hauptmann and some of the men pointing guns at me; he beckoned me, and laughingly said, 'Ah, so, you haff come to us!' Apparently they had been summoned to search for me.

This time I was taken to the station police, station headquarters, and handed over to them, placed between two burly tommy gun armed soldiers. Later I was put into a small open truck with four of them guarding me. They delivered me to the Gare du Nord, but I did not see my old guard again; there I was, put on a train for Germany, with two guards this time. Crossing Paris it seemed very like London; lots of civilians, some military but not warlike really. On the train I slept most of the time, by now worn out and miserable, still in the battledress that I had ditched in, dirty, creased, but at least dry. I was given some army rations, I remember, but have little other recollection of the final journey, feeling despondent ... and a fool.

The next stop was in Frankfurt-am-Main; there I was handed over in the station police station to two Luftwaffe NCO's and driven in a sort of van to Dulag Luft, the Luftwaffe reception camp for RAF and Fleet Air Arm prisoners. I had a vague recollection of a brief lecture about Dulag whilst with 43 Squadron and about giving the usual name, rank and number only.

Arriving there, I was initially put in a reasonable room in the main building and interrogated by several English-speaking, smooth and persuasive interrogators. Giving my name, rank and number, I ignored the obviously phoney large Red Cross form given me. Later I did fill it in with my name, rank and number only anyway.

The next day was quiet and through my window I watched other prisoners in a sort of parade ground outside, surrounded by the barbed wire fencing which was to become a familiar 'hedgerow'.

Another interrogator came in, well spoken – he was just too bloody good to be true. We got nowhere I think and he left soon after. This, I learnt later, was the notorious Eberhard who had spent considerable time in England; a truly nauseating specimen.

A couple of days later I was transferred to the camp proper, wooden huts. There I was quizzed quite stiffly by their reception committee. They seemed a queer lot to me, RAF and Fleet Air Arm, and there was a most peculiar atmosphere about the place which was very off-putting. I was warned that the rooms were bugged, not to trust any 'goon', their name for the staff. The whole set-up did not help a newcomer to get POW orientated; in fact after a few days I began to wonder just what the hell was going on. We learnt a lot about this place of course in later years, some good and some bad ... enough said!

Within a week I was told to get ready to move out to another camp. Get ready, that was a laugh. With what? The clothes I stood up in plus one blanket, one Polish greatcoat and a couple of books! We entrained, about 100 of us, and were taken to Lübeck on the Baltic; at least we were taken in passenger carriages this time, not cattle trucks as later on. At Lübeck we were marched some distance to what looked like a large camp. It turned out to be the most miserable, inadequate, badly staffed camp possible, with low morale and again a funny atmosphere. Here, we mixed with Army officers as well. It was getting into winter weather and everyone was cold and underfed.

It was a camp of wooden barrack blocks with the usual double barbed wire fencing round the perimeter, with the internal trip wire over which one did not venture without the risk of being shot at by the guards in the lookout towers situated at strategic points along the wire fencing. They had machine guns and hand grenades at their disposal, with searchlights for sweeping the fence and camp at night. Accommodation was primitive: the standard two-tier wooden bunks with straw-filled mattresses and pillows on wooden slats; the huts overcrowded, stuffy at night from the terrible coke stoves and closed windows.

The atmosphere was spiritless, especially amongst the Army prisoners, most of whom had been taken in France early on. Escaping or any such organisation seemed to be on a low priority and was not discussed with us 'new' boys who were always treated with suspicion at first, and interrogated on this premise. Amongst the RAF and Fleet Air Arm the morale seemed better, being in the main younger and newer, but the camp was not organised as subsequent 'hostelries' were in later years.

The food was a near starvation diet; four or five potatoes, some weak soup, four slices of their army ration brown bread, about a quarter inch thick with ersatz butter and jam, ersatz coffee and saccharine. A very small ration of sugar was issued. There were no

Red Cross parcels available at Lübeck, or at least none being issued.

The staff and guards were a scruffy lot and arrogant with it. By this time, October 1941, on the Baltic coast it was getting cold, especially at night, but the days were generally fine and quite sunny. I joined eighteen others in a large room with the bunk beds arranged round the walls, listened to their chat and experiences, and endeavoured to adjust to the life. One highlight was the rewarding efforts to entice a large black cat into our compound from the adjoining German compound; after several days of encouragement it finally came through the wire. Once there is was caught, killed and skinned, the flesh eaten by those who had planned it and the skin used by another who won it, to make a pair of mittens out of, in backwoodsman style.

The next excitement was a night raid by the RAF on Lübeck when they dropped some bombs near the camp, hitting the German compound and also the sick bay, or Lazarette. Unfortunately, an incendiary went through the leg of a British inmate. Naturally we cheered the hits and fires on the German compound, until threatened with being shot.

After several weeks at Lübeck, in which I kept very fit and tried to learn as much about POW life as possible, we were told that the camp was required for others and that we were to be transferred to a camp at Warburg, near Kassel. We prepared for the move, getting together our meagre belongings. By now, I had acquired a sort of kit bag and an army side cap with ear muffs from a bulk issue, some paper, an exercise book, pencil and a pair of army boots. In the cold, most of us who still had them wore our flying boots which proved to be very strong and serviceable over later years, a good advertisement for the maker, whoever he was!

The most annoying aspect of camp life was probably the frequent '*Appells*', parades for counting heads, which usually provoked some incident or other between the Senior British Officer and the Germans, or just between us and the guards doing the counting, as we shuffled about and made it difficult for them. The final count before leaving for the train for Warburg was a long drawn out affair as they were checking each prisoner's belongings, number of blankets and identification. There was one ridiculous scene when an Army padre, a fellow of about 6 feet 3 inches tall, thin as a rake and very garrulous, was found with two blankets – a heinous crime! When called upon to give one of them up, he became almost hysterical and the guards grabbed one of them, he grabbed it back and a tug of war ensued, watched and cheered by the rest of us. Finally he was overpowered and the blanket taken, but it was an illustration of the state to which prison of war existence can reduce even the most outwardly stable people, and the pettiness of it all.

We were marched to the railway station and entrained in a long passenger train with guards on the roof, a 'flat top' waggon at the rear with a machine gun and about eight guards on it. Entraining was carried out in an orderly fashion under the eyes of about a hundred armed guards, no chances being taken. All had received some rations for the journey and we settled down, keeping a look-out en route for anything worth noting, with patrolling guards in the carriages.

It was an uneventful journey, with several stops on the way. No one made any escape attempts as there were few opportunities. Arriving at Warburg, (Oflag 6B), we were kept outside the camp proper for several hours; the waiting was tedious.

Once inside, blocks and rooms were decided upon and settled into whilst the more experienced 'X' chaps looked round for the best positions and available facilities for their various activities, with the emphasis on tunnelling sites.

It was a large camp covering an extensive area, without trees of course, and the surrounds well cleared of obstructions that could provide any cover. At one end was a long red brick building, the rest being large wooden barrack blocks on raised floors, but with concealed space under the flooring. There was ample room for sports grounds and gardens and a communal dining hall adjacent to the cookhouse.

The barrack blocks were quite far apart so that each became a sort of separate entity, split into rooms taking eighteen to twenty people, with a small room at one end for two only, the block commander and his adjutant, opposite which was the night latrine, a room with several large buckets only! Accommodation was crowded and each inmate stored his small cache of belongings on, below his bunk, or in a locker.

The camp had a large but very mixed crowd of prisoners; British Army, Royal Air Force, Navy and Fleet Air Arm and some Indian Army Officers. Various committees were set up to organise and co-ordinate all activities and it soon assumed the guise of a well run township. Initially, priority was given to setting up sports and gardening facilities and get them going, with various instructional classes as well. This was to lull the Germans into a sense of tranquillity, accepting our 'peaceful' occupations and seeing great activity about the place. However, behind this screen the serious work of the Escape Committee, the 'Big X' organisation, was gearing up.

With the very many experienced prisoners now here, and more arriving, there was a wealth of talent and the organisation was well run, efficient and resourceful. The educational classes got under way, using the vast range of qualifications and skills available; these were also a superb camouflage for the more serious endeavours.

As a 'new' boy, it took some time to be accepted and also to adapt to the life; after a while I joined the team on a tunnel from one of the latrines, my basic engineering training being useful. This was an 'Army' tunnel and did not last long, being discovered after some weeks' work. Unfortunately, it alerted the previously rather naive guards and security was tightened up inside the camp, but it was still inefficient.

The winter of 1941 was severe and the cold intense. At night we were more or less locked in with closed windows and the fug with red hot coke stoves was abominable, to be relieved in the morning by a short, sharp run and naked roll in the snow. Really invigorating! The three Services maintained a dignified individuality and identity, but co-operated where necessary. The Navy and Fleet Air Arm boys, in our block, were standoffish and aloof. The Indian officers lived almost separately, but were a wonderful crowd; on parades they were always the smartest and we marvelled at the way they maintained their clothes. They were a well decorated lot, from the First World War and when invited to take tea with them it always began with a silent toast to the 'King Emperor', most impressive, with all standing to attention.

At Warburg, I experienced my first Red Cross parcel. These were issued individually, possibly pooled in each room under a duty cook for the main items, but we still had a daily meal in the communal dining hall; usually spuds and a weak soup which we would augment by adding a Marmite or Oxo cube, a great treat initially.

The gardens took on a spring-like appearance and it was decided to use this as camouflage for the air holes necessary for a long tunnel from our block. By now several tunnels had been found or abandoned for various reasons, but a major one was started from our block. This was a long way from the wire fence and was unlikely to attract any attention for this reason. Planned to pass underneath the end of another block closer to the wire, it was to be about 200 yards long, eventually coming up well clear of the perimeter near a barn. It was thought that it might even be possible to reach the barn and come out in it so that a regular trickle of escapees might be possible without giving the tunnel away, or it being discovered. This was an ambitious plan.

In the event, the tunnel was made with a false entrance and a dummy tunnel to fox the 'ferrets', as we called the guards who roamed the camps looking for evidence of tunnelling – or anything else. First, a suitable trap was made, then a shaft dropped, a short one, and a tunnel dug for about twenty feet with another trap door in the floor into a second shaft going down a further six feet. There the working chamber was made, a room of about five feet square, to store tools, waggons, equipment and clothing. All this was carried out in heavy clay which was very difficult to work.

From the workroom a second tunnel was dug, electric lighting installed and a light railway line of wood laid to cart the residual clay back to the workroom to be stored for dispersal. The electric lighting was a single wire system using the ground as one earth return, the wire and fittings scrounged from every possible source in the camp. This system was dangerous and in fact in a similar tunnel dug from the dining hall, a digger was nearly electrocuted and only rescued and saved by calling the 'ferrets' and giving the tunnel away. We had to be very careful.

With the dummy tunnel now only used as an entrance to the real one, progress was quite rapid; the heavy clay being dispersed in the extensive space under the floor which had brick walls and was not inspected by the ferrets, as Warburg was not yet fully conversant with 'Kriegie' activities in this field.

The clay was cold and damp and tricky; we tried to do without shoring the sides and roof of the tunnel, but soon found it necessary. For this purpose the 'bedboards' were ready-made for the job; everyone subscribed until few were left and had to be replaced by hammock style beds made up from the strong string binding of each Red Cross parcel. Once, when down on line maintenance, I was nearly buried under about a ton of clay which fell out of the roof in one enormous lump, just managing to pull back in time. This lump was a wad of solid clay shaped like a dome and left an inclined hole in the tunnel, but the right size, over the lump as it were. As time was important, and the lump so large we had to quickly shore the area above it and make another junction in the railway as the line and little waggons could not negotiate the incline. It also meant having two more men in the tunnel to get the waggons over the hump.

The main tunnel was now twelve feet down and air shafts had to be cut up to the surface for ventilation. Making 'cutters' from suitable ex food tins, we made extending rods of wood and bored upwards with the cut-out clay falling down through the tins. Using different sized tins the holes were tapered towards the top; there they were carefullly disguised in the foliage of the now fertile vegetable gardens, watched and inspected by 'gardeners' for subsidence, and covered with a stone when not required.

The object of the dummy tunnel was, of course, that if discovered the Germans would think that we had abandoned it because it was too far from the wire; in this event the first leg could then be used to disperse the rest of the displaced clay in. In actual fact this is what happened; they discovered it, made much of the stupidity of attempting such a project, destroyed the trap and left it.

As soon as possible, a third shaft was sunk from the end room in the hut nearer the wire, straight into the tunnel. After several more weeks it was found necessary, for various reasons, to go for an early 'break' before reaching the planned exit point.

On the night the break was made, the first to emerge were immediately picked up by an outer patrol, which had been put on earlier than expected. We had been warned of this additional patrol through a 'tame goon' contact. Panic ensued and the sixty or so escapees positioned in the tunnel ready to go had to back track as fast as possible. There was some wild shooting into the camp for a while, no damage, as the guards ran in not knowing what was afoot. In the chaos the majority of the escapees managed to get back, up into rooms and disperse their escape equipment and rations as best they could and avoid detection.

After several days' constant parades, head counting, recriminations, examination and searching of rooms, gardens and final flooding of the tunnel, we settled down to a peaceful existence. It had been a gallant effort, kept the Germans occupied, given all concerned an incentive and occupation during the worst months of the year.

This long tunnel was only made possible by the fact that we could disperse the clay removed in the under-floor space even in the winter, but never again once discovered!

The flooded tunnel was not all filled in and there was a twelve or fourteen foot length left open between the trip wire and the main perimeter fencing and two of us sought permission to try a blitz tunnel from it, Dennis Graham-Hogg and myself.

The idea was to drop into it with two light brown blankets, crawl to the outer fencing, hide under the blankets until dark and then tunnel a shallow ditch under the wire, a matter of about sixteen feet or less, to come up just outside. Being hard against the fencing wire, we were effectively out of sight of the tower guards and in the blind spot made by the line of fence poles and wire. Organising a fairly large band of perimeter walkers, we positioned ourselves in the middle, whilst two diversions took the attention of the nearest tower guards. Timing our 'walk' to reach the open remains of the tunnel when the outer sentry was at the furthest end of his beat, we dropped down into the tunnel and crawled to the end against the wire and hid under our blankets. Kept informed by friendly shouts by 'walkers' we lay there for half an hour or so, but were warned that a sentry had stopped opposite us and was looking into the tunnel remains. We tried a quick, careful look from under our blankets, but were unable to see him over the tunnel edge and the coiled wire in the middle of the two lines of outer fencing.

Getting worried we could not move and it was getting darker when our watchers would have to go indoors, two 'walkers' sauntered past and confirmed he was still looking and warned us to be ready to come back. Apparently the guard was getting agitated; we were told to be ready to 'rejoin' as a diversion was laid on nearby. This was a mock fight on the perimeter track; a band of fellows walked by and we got up and ran to join them. The guard was now confused, as he knew something was happening but could not see what.

He blew a whistle and we dispersed quietly, leaving behind one blanket which was, of course, discovered the following morning. Then we saw that we had disturbed the ground when dropping in and this, we thought, must have caught the sentry's eye. In retrospect it was an impulsive attempt, which would probably have got us shot out of hand anyway as we were over the trip wire, and we had not really given it any planning.

We had several Ulstermen in the camp by now; New Zealanders, Australians, Poles, Czechs and others. Sport was being encouraged; a skating rink had been improvised by banking up round a football pitch and flooding it; the icy conditions had done the rest. With skates bought from the Germans, some of the Canadians soon got ice hockey going on a small scale. The camp theatre was by now in full swing, with great talent from the Army officers and some from the rest. Several orchestras and small bands were in the making, with some instruments supplied and bought from the Germans and others sent from Switzerland and other places.

The various 'X' organisations were running smoothly and the coded messages filtering back and forward from the UK in the mail had improved since Lübeck; despite the run of the war in early 1942 we were optimistic and the entry of America into the war was welcomed, not the least for its enormous manufacturing potential and logistic support. The Germans were still very cocky and the guards at Warburg – the dregs of their army, we reckoned, except for a few – were being bribed to bring in many necessary, but forbidden, items, plus a good supply of eggs!

In 1942, another valiant and successful escape was made, this time going 'over the wire' in a unique manner, using built-up extending ladders the upper half of which dropped across the double perimeter wire fences. The bottom ladder was placed against the

Author flying Spitfire MK XVI at RAF Chivenor, 1947

Author flying Spitfire MK XVI, RAF Chivenor, 203 AFS, 1947.

Author with Flight Lieutenant Maguire and Flight Lieutenant G. Mott, DFC. Gunnery Squadron RAF Chivenor, 1948.

Author Commanding Gunnery Training Squadron, RAF Chivenor, 1948

Fighter Command Rugby Team 1950–1951. Author fourth from left, back row. Captain – Squadron Leader Horsfall

Author with instructors and staff of the Flying Wing, 226 Fighter OCU RAF Stradishall, 1952–53. Aircraft is a Meteor MK8 with which the unit was equipped.

Author flying Hurricanes MKIIC, No. 363 (now on the B of B flight) at RAF Stradishall 1953.

Squadron formation of Meteor MK8's flown by some of the Flying Wing Instructors of 226 Meteor OCU RAF Stradishall, the author leading, 1953. The Flying Wing staff could 'scramble' 24 Meteors in 2½ minutes in the fighter role from ASP readiness.

inner wire and the top one pushed up by the first up the ladder, to fall across the gap, to make a bridge over which the escapees crawled to drop outside the wire to freedom. The main snag was that to be successful it had to be done at night and the perimeter lights and tower searchlights had to be neutralised. This was to be done by fusing the entire camp lighting system by throwing a grappling iron over the overhead supply lines.

After making the ladders to the required size from timbers scrounged from roofing trusses, under-floor joists and other places, teams were selected and practice runs made in the huts after dark, or during suitable secure periods during the day.

On the night selected, with teams in position, the lights were very successfully extinguished completely, as though doused during an air raid. The scheme worked like a dream, but our ladder unfortunately tipped up and into the middle of the double fencing. The 'number one' had run up too close to his bridge, he fell back into the camp and we all dispersed. The rest, three ladders, made it and some twenty-four or so got over and away.

There was pandemonium, guards firing wildly into the camp, others firing outside, only one of their own men being hit. They were really scared this time, in darkness and not knowing what was going on, by the time they got themselves organised our successful efforts, except for the ladder, had been well hidden and we assumed an innocent posture, during the trying days that followed. Those who got away, however, soon arrived back in penny numbers, none making it out of Germany. The nuisance value was terrific and as a morale booster it did wonders; our relations with our captors degenerated and life was made difficult by reprisals, but all well worth the inconvenience. Visits by the Gestapo, heartily disliked by the German Army, to their credit perhaps, and some 'SS', the Kommandant and his staff taking the brunt of it all, but in their funny way accepting that we were doing our duty and respecting us for it.

I, personally, had my first taste of the cooler, solitary confinement, for striking a guard under provocation. The stoves in our rooms were coke and coal burning and we collected our ration, in this instance of coal, from the red brick building at the far, or top, end of the camp. We would carry up our bunker box between two of us and queue with some forty others outside a sort of wire cage in the basement, entering one at a time to fill the box under the watchful eye of a guard. When getting coal, the large lumps were on the top and great quantities of dust below; naturally we went for the lumps, to the anger of the guard who insisted on us taking dust as well.

In my case I stubbornly refused to do so and went on taking lumps, the guard was getting exasperated and when bending down he hit me on the back with his rifle butt. I elbowed him in the groin and chopped him as he fell forward. Loud cheers from those queueing; the other guards entered the cage, clobbered me, but not too hard and one of them marched me off at rifle point down through the main camp, followed by a jeering crowd of prisoners. Nearing the block in which the Senior British Officer, Wing Commander 'Hetty' Hyde, lived, I veered towards it only to hear a round being snapped into the breech of the rifle about three feet behind me; then it was jabbed in my back. I kept going straight, but called out what had happened and was given a 'thumbs up'.

Through the main gate and into the cooler on the immediate left on the outside in the German Compound, into a cell to cool my heels. It was in fact larger and better furnished than the one at Lannion airfield. By standing on the table I could see out of the small window. The camp people sent me some books, paper and pencil and some Red Cross chocolate and biscuits. This bit of solitary was in fact ironical and farcical, as after a few days the guard who had struck me first was brought in for a short spell himself, after a brief investigation into the circumstances and evidence by those watching. One evening the guard commander let me out to enjoy a beer with them. I learnt much about the goings and comings through the main gate by watching and timing them; this was useful to the 'X' boys for future use.

After two weeks I was released, well rested and quite well fed with the extra Red Cross bits that had been sent in.

Warburg camp was the first to introduce an 'Exchange and Mart'. With regular Red Cross and Personal parcels coming through, the Red Cross prisoners could afford to be more selective and exchange articles. The Mart operated on a 'points system', each article being priced on points, a percentage being retained towards a pool of escape rations. We also had a bad epidemic of diphtheria to which I succumbed and ended up in the sick quarters where we were looked after by one of our own Army doctors, a mad fellow who had been taken early in France. Doing his rounds he brought his imaginary dog with him, inviting it onto the bed of each patient, talking to it and the patient was supposed to do likewise – and pat it!!

The supply of serum 'anti-dip' had not arrived and we were in rather a bad way, about forty of us. When it turned up the doctor, actually a splendid chap, approached me with a syringe the size of a grease gun, with a needle like a six inch nail, which he plunged into my left thigh. Giving four savage pumps he unscrewed the syringe from the needle, which he left in my thigh, and went away to refill the syringe, saying, 'I'll be back in a minute'. Three quarters of an hour later, still with this perishing needle in my thigh, he came back, saying 'Sorry old chap, old friend of mine just arrived, had to give him a cup of tea'! Screwed on the syringe once more and pumped away, this time taking away both the syringe and the needle. I recovered soon after.

This was my first real POW camp on a large scale with the most fantastic collection of men imaginable, from literally every occupation you can think of, every walk of life and background, from barristers to crooks. They brought with them a wealth of knowledge, tradesmen, teachers, professional men, career service men, engineers, clerks, technical experts. Their contribution to camp life, the escape and intelligence organisations, entertainment and general affairs would be the subject of a book in itself. All these things required enormous organisational effort and the full co-operation of everyone, whether ardent escapees or not. There were many who would help in every possible way, but who were not motivated to get out.

The camp assumed all the trappings of a small and highly organised town, albeit under great stress, but at least no one was going to go missing for long!

Towards the end of 1941 we heard that we were going to be segregated into separate camps for each Service, administered and guarded by similar German Services and that the RAF prisoners at Warburg were going to Schubin, in Poland, not far inland from Danzig. This was welcomed, as the chances of escaping, possibly to Sweden, looked promising.

CHAPTER THIRTEEN

Schubin (Offlag XXIB)

We were moved in well-guarded cattle trucks to Schubin, some seventy miles from Danzig, near Schubin village and a railway station. Upon arrival we had a weary march to the camp. It was totally different from Warburg or Lübeck, on the slopes of cultivated agricultural land with trees nearby; the camp on one side of the road and the main German Compound on the other, opposite the camp. There was one initial snag; we had expected to be arriving in a Luftwaffe-controlled camp as informed, but no, it was another Wehrmacht administered camp ... so much for the segregation under similar Services!

There were usual perimeter fencing and watch towers, some of which had British Bren guns, captured in France, as the Germans found them accurate in short bursts and also with not many available this was an efficient way to distribute them. There was a large white mansion house, later learnt to have been an orphanage, dominating the other extensive buildings and barnlike barrack blocks. We marched in and paraded inside the wire, forraging parties quickly carrying out a reconnaissance of all the buildings, selecting strategic places for the various activities. After we had reported back to the Senior British Officer, Wing Commander, 'Hetty' Hyde, it was obvious that the place was uninhabitable. It was filthy everywhere; broken furniture, beds crawling with bugs; dirty, broken straw-filled mattresses, the floors covered with garbage, the smell and stench of stale urine and filth abominable. The SBO inspected it all himself with his staff and ordered us to remain outside whilst he went to see the Kommandant, a stiff Prussian, much older with an artificial leg and an arrogant manner. While the SBO was in the German Compound several of us went scrounging, looking for good sites and, with two others, I was picked up by guards in the block nearest the road. We were taken across to the cooler without being searched or questioned, but put inside, in fact, with several useful implements on us. The Germans at Schubin had not yet experienced British prisoners!

The cooler was in an unfenced field with some trees not far away, the window was mid-height, quite large and of course barred on the inside. That evening I was able to saw through three of the bars, seal the cuts with a mixture of spittle and dust and get some sleep. We had not been fed, or given a drink. Next morning we were released, taken back over the road and into the camp. Reporting our findings to the 'X' boys and suggesting that they get someone into the cooler with a view to slipping through the cut bars, we rejoined the rest. It seemed that the SBO had indeed convinced the Kommandant that the accommodation was uninhabitable and he had agreed to bring in some Polish workers to clean it up. Later a hoard of Polish women did so and we moved in the next day; we heard that the Kommandant had gratuitously agreed that the camp had been occupied by French prisoners and that 'they were not a clean race'!

Regarding the cleaners, it seemed that most of the able-bodied men above sixteen and under fifty had been conscripted and taken away to work, as we never did see any other than women and very young or old men.

Once we got into the buildings they were of necessity divided into rooms or compartments using the tall, narrow lockers as partitions. There was no privacy and the noise and fug when locked in at night was almost unbearable. The mood in the camp was dangerous from the start and morale low, but everyone did their best to settle in. The communal lavatory was a long forty-seater, straddling a foul-smelling pit which was periodically emptied by a 'Honey-Waggon' horse-drawn wooden tanker with an antiquated hand pump to fill it with; it was foul smelling and insanitary. The washhouses were bare, concrete walled and filthy. It was not encouraging.

Soon after we arrived, tragedy struck when a chap, who was in rather a depressed state and when walking out was usually accompanied by a room mate, decided to just walk straight to the wire and commence climbing it from the inside. Immediately several fluent German-speakers contacted the watch tower guards and told them he was 'Krank', sick, and to let him get over and pick him up, or let his fellows cross the trip wire and try to persuade him to come down. However, a foot guard outside refused to listen, panicked and shot him at point blank range. By this time a hostile crowd had gathered and the situation was ugly; there was nearly a mutiny as he lay on the ground and no one was allowed near him, extra guards with tommy guns were brought in whilst he was removed, only to die later.

Several days later an RAF Sergeant escaped to Sweden. Sergeant P. Waring, together with others, was on a working party to fetch bread from the nearby railway sidings when a train bound for Sweden halted near by. On impulse he got up onto a bogie and in a matter of hours was on the train ferry, into Sweden and back in the UK four or five days later.

As the mood in the camp quietened down, we were suddenly to receive a whole new contingent of RAF prisoners from Sagan, including some of the longest serving and most experienced escapees and 'X' men. Accompanied by a, later to be liked and respected, Luftwaffe Lager Feldwebel Hermann Glemnitz, they were handed over to the Wehrmacht in no mood for nonsense as they too had expected to come under the Luftwaffe still, as they had been at Sagan.

These new arrivals were a welcome sight. Wing Commander 'Wings' Day took over as SBO and the organisation of the camp took on a new look under the guidance of these experienced operators including Jimmy Buckley, Aidan Crawley and others well versed in the escape and intelligence requirements; a radio was built from parts smuggled in from Sagan and life took on a new dimension.

They wanted to try the Warburg 'over the wire' method so we made ladders, the greenhouse, a proper one, providing suitable camouflage for constructing them in, as additional shelving of course, if suspected! The mood was tense and all efforts were mobilised. The attempt to emulate Warburg was, however, doomed, for on fusing the lights by the same means, rushing the ladders to the wire and erecting them, the lights suddenly came on

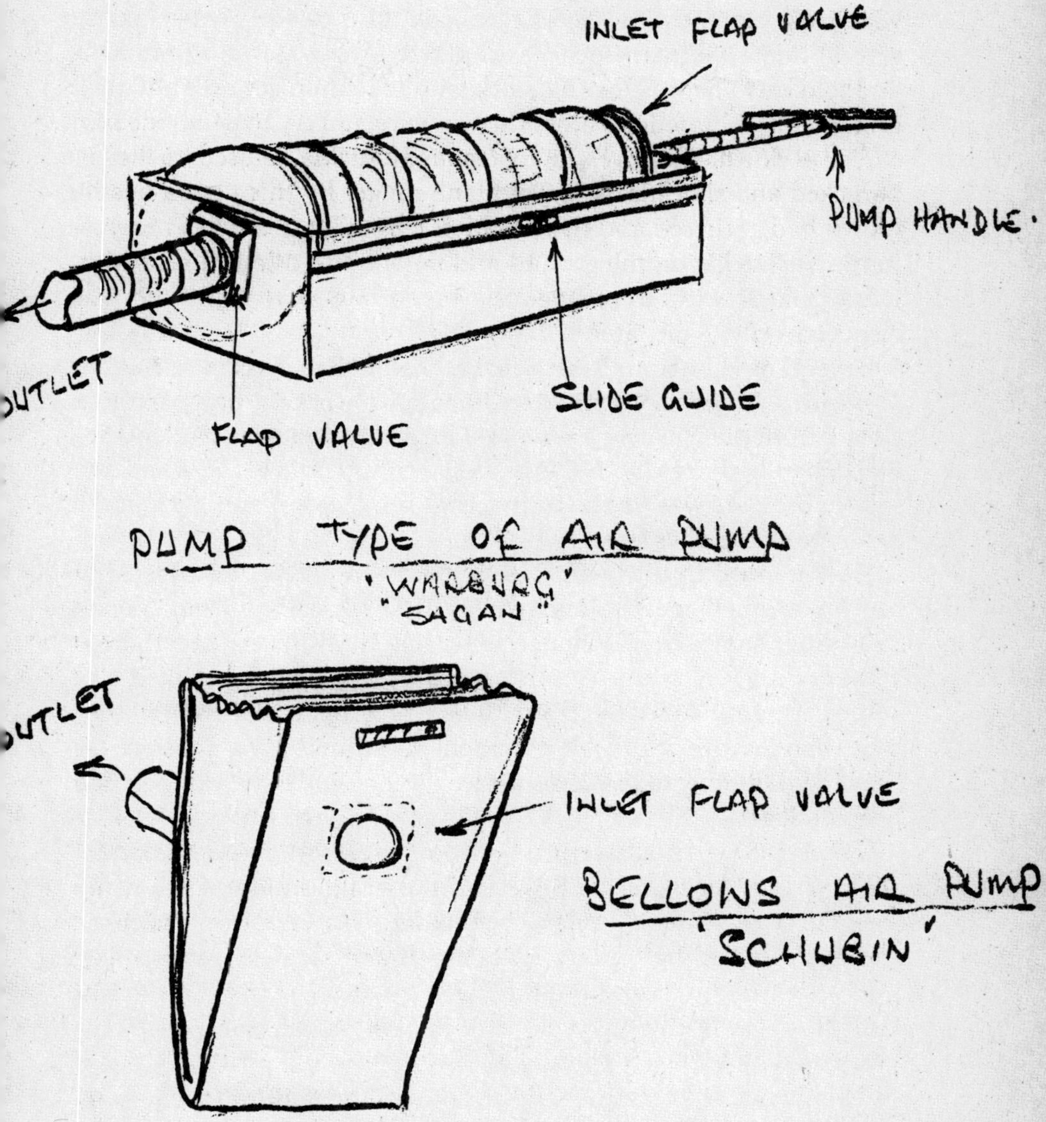

Air pumping equipment used in the Schubin tunnel escape

again, or a reserve set, operated by an emergency system. There was chaos of course, followed by recriminations, reprisals, changes of guard systems and alterations to the wiring to eliminate any recurrence.

The atmosphere between us and the Germans worsened and the SBO's liaison with the Kommandant deteriorated through no fault of his. The heat was turned on to give the camp staff maximum noise value; tunnelling was intensified, with five going at one time. It was 'all systems *go*', but security against the guards was difficult as they patrolled constantly inside and out, no 'ferrets' as such, but very inhibiting. Additionally there were hidden microphones in the sandy soil round the perimeter to detect digging. The winter of 1942 was intensely cold and the fug in the barrack blocks froze on the *inside* of the windows and the walls; in this cold, many of the activities were reduced.

The New Year came and tensions eased as the workload increased, but as the camp buildings were closer together than hitherto, the 'stooging' or watching and following the wandering guards became a difficult task and several close encounters took place.

One tunnel entrance was from the wall of the communal lavatory, the entrance being through a lavatory seat with the air inlet to the pump situated near by. When a shift was down this seat was maintained 'occupied' by details of 'sitters' with their pants dropped, squatting with a book. The air inlet pipe, made from round food tins fitted and sealed into another, amplified the noise of the 'flap valve' as it opened and shut on each stroke of the kit-bag pump. One day a newly arrived prisoner squatted on the adjoining seat by the watcher, on vaguely hearing the sound of the flap valve and noting the apparent deep concentration of the watcher, he got up and went to the MO thinking that the man was ill! Telling the MO, our Army doctor from Warburg, about it, he was shocked when the MO burst out laughing. The doctor enlightened the new arrival!

After the hard winter every effort went into the several tunnels, the one I was working on from the washhouse was slowing up due to the central position and the extra guards around, but the Abort, or latrine, one was going ahead well.

An oft-mooted plan to use the 'Honey Waggon' was now agreed, with co-operation of the driver, who had established close liaison with a Czech prisoner, known as Ricks, and the Polish resistance outside. The idea was to hide two prisoners in the tank of the waggon, suitable clothed in overalls to take the smell, etc., whilst the driver promised to clean the tank out also, and drive out of the camp as usual.

With the progress of the Abort tunnel, it was essential to co-ordinate the 'Honey Waggon' escape with the 'Abort' one so as to eliminate any chance that one might jeopardise the other, and not compromise the waggon driver.

Whilst this was being organised the SBO was informed that the camp was to be evacuated and we were going to be moved once more in the spring; March was quoted. It was decided that to give these two efforts maximum security and the chance of success they deserved the other tunnels should be suspended for the time being. This was done and on March 3rd all was set; the tunnel was packed with the chosen escapees with their kit, as they had to be shut down inside before the barrack blocks were locked for the night. The Honey Waggon had already driven through the main gate without being checked. It never was; well, not inside the tank anyhow, with Ricks and 'Boss' Morris inside it.

When morning came, we found that thirty-three had got clean away from the Abort tunnel and the exit had not been discovered. This was unbelievable and Don Gericke immediately gathered together his escape kit and was away down the tunnel and out before you could say 'Hitler'. That made it thirty-four plus the two in the waggon, a magnificent effort. Joe Ricks was not seen for several months until he rejoined us after we had been moved, but he reached Warsaw where he spent three months until retaken by the Gestapo, who gave him a very rough time indeed. None of those who got out managed to reach home, though several were free for up to two weeks. This escape involved over 4,000 troops being moved into the area to assist 1,000 or more German police and homeguards in searching for the escapees. Truly a successful operation, despite the disappointments.

Meantime, as the panic and usual reprisals and chaos subsided, we learnt that the Kommandant and the Abwehr Officer had been suspended and new ones appointed; Rittmeister Reimer was made

Security Officer, a decent, cheery fellow, who carried out his duties effectively but reasonably.

Under the new Kommandant the Ferret system was introduced, deep ditches were dug just inside the wire and small explosive devices were detonated round the perimeter to discourage tunnellers. Although the other tunnels had been kept inactive but intact for a while after the Abort escape, when recommenced the tunnellers were beaten by the security measures before another break could be attempted.

At the end of April 1942, the whole camp was paraded and told that they were to prepare for another move, this time to Sagan, Stalag Luft 3, a Luftwaffe-administered establishment. Feldwebel Glemnitz arrived to supervise the loading of our six hundred strong contingent. Checking the railway arrangements he found that we were again going to be transported in cattle trucks. He was livid, because it was too difficult to keep a check on prisoners in these trucks and escapes would be inevitable as they cut their way out, as had been done previously. He succeeded in having the trucks replaced by third class passenger coaches, so we travelled in reasonable comfort, but with little chance of getting out on the way.

I had now been a prisoner for one year and seven months, in four different camps and worked on three tunnels and two 'ladder' jobs, the second time in a minor role of construction only. Settled into the Kriegie life I had made few friends and remained solidly independent. Letters and periodic parcels from England relieved the monotony, and the many projects in hand, together with sport and reading, kept me well occupied. However, I was beginning to query the chances of getting away before it all ended and was almost convinced that to 'make it' one had to have an element of luck, be able to speak German and be exceptionally well prepared. Personally I could not speak the language and felt that my luck had run out, for the time being anyway. Therefore, I more or less resigned myself to assisting in every way that I could, but was indifferent to actually getting out.

At Warburg I had, like many others, constructed myself a box for my personal belongings. This was made, not with the usual false bottom, but with a false back with a row of sharp nails protruding inside to deter the ferrets from measuring the dimensions with their hands and arms. With a space 2½ x 24 x 18 inches available, the false back was used to carry all sorts of 'goodies', master maps, tools, compasses, a camera, radio components and other bits and pieces. This false back was never found; having the hinges three inches forward of the back edge they never suspected it. Carried by two people, the handles were positioned so that it balanced evenly despite the additional concentrated weight at the back. It had survived Warburg and Schubin and now arrived safely at Sagan.

CHAPTER FOURTEEN

Sagan (Stalag Luft 3)

Arriving at Sagan was an experience in itself. Here was a large purpose built, supposedly escape-proof. POW camp administered and run entirely for the RAF prisoners by their opposite number, the Luftwaffe.

Built in a clearing of about sixty to seventy acres in a forest of small pine trees, the camp was planned for expansion and had already been enlarged. There was the original East Compound, the new North Compound and the German Compound and the Kommandanture separating them with a Lazarette (Sick Quarters), cooler, guard rooms and barracks in an adjacent *Vorlager*; to the south and west the site had already been cleared for expansion and the whole area was well clear of the surrounding pine trees.

Located about eighty miles south-east of Berlin, with a convenient railway station nearby at Sagan township, we could both see and hear the RAF bombing raids on Berlin under certain weather conditions. Indeed, later on when these raids increased and 'block buster' bombs were used, some claimed to actually feel the detonations. The North Compound was new and almost a model POW camp. This was partially occupied by a party of about two hundred and fifty RAF prisoners, some of whom had initially volunteered to go across to help get the camp ready. These stalwarts of long 'X' operations had already made their plans for tunnelling and when finally transferred from the East Compound soon got at least three going. So when we arrived and were split up between the North and East Compounds, those who went to the North found the tunnelling and escape organisation a going concern run by many of their old friends, including the SBO.

I went with the other party into the East Compound. This was an intensive area, well clear of the trees; large, well separated wooden barrack blocks raised on short stilts to enable the Ferrets to crawl under the floors and search for dispersed sand, or signs of a trap. The distances to the wire were prohibitive and the soil was sandy and unstable so that full shoring was essential for tunnelling. There were two fire tanks, in fact pools dug in the ground which could be used for swimming in and the recreation facilities were plentiful. The accommodation was better and clean and the efficiency of the camp vastly superior to anything we had experienced previously.

On arrival some of us established a half 'Ulster' room; not quite enough for a rugby team. Again eighteen to a room and the usual arrangements. The Red Cross parcels distribution was assisted by our own people in order that certain marked parcels containing 'contraband' could be extracted before opening or handling by the Germans.

The food was good, with Canadian and Argentine bulk issue parcels augmenting the standard English parcels, and the 'Mart' did a roaring trade. Morale rose and despite the poor escaping prospects, the many other activities flourished and were expanded. German speakers made wide contacts with guards and Ferrets which provided much valuable information. This, combined with interrrogation of the many new arrivals (who noted things wherever they were captured or taken), was sifted and sent back to Britain via the coded messages which were now highly organised.

Relations with the camp guards, ferrets and administrative personnel was on a more common ground, being both Air Force, and additionally the war was not looking too good for Germany, so many of them were looking towards their own interests; or simply to get cigarettes, chocolate or soap. The wide educational activities were expanded and the 'Sagan University' got into its stride with books arriving from the UK and other sources.

Resisting the temptation to relax, we decided to tunnel from the cook-house, the nearest building to the wire on the east boundary. This had a good sized potato store opening onto the wire side with a large recreation hall behind it, both covered by a roof with a large area of ceiling made of tongued and grooved planking.

The concrete flooring once broached made a tight fitting trap surround and we cast a trap of matching concrete which fitted beautifully, the cracks being filled with matching grey dust and potato dirt. The sand proved easy to work, quick to dig, but difficult to shore, the displaced sand being dispersed in the roof above and spread evenly. However, we miscalculated the volume required to hide the sand in, because once disturbed it took up considerably more space than when well packed. This was to be our undoing as the weight of the greater volume dispersed caused the roof to sag suddenly one morning, and some sand spilled into the classroom below. We had then gone about thirty yards in record time and tried to move the layer in the roof; it did not work and a couple of days later one of the ferrets found it. As we had closed the trap down securely it took them a day to find it; having done so they brought in the fire engine and flooded the tunnel. They would never go down into one. We were dismayed but not downhearted; it had been one of the cleanest jobs ever and set the scene for a future successful escape from nearby.

This was the famous 'Wooden Horse' conceived by Williams, Codnor and Philpott. The story is too well-known to be repeated, being so well presented in William's book and subsequent film. I helped with the design of the horse, but had no other involvement. It was a magnificent effort.

At the time their Wooden Horse was going well towards the same east wire that our tunnel had been aimed at, I displaced a cartilage in my right knee in a big way, playing rugby in a combined Irish, Aussie and New Zealand team versus the rest. It was a particularly hard tackle by Squadron Leader 'Dim' Strong that did it, catching me turning on the right leg only. Despite the MO's attempt to straighten the leg by jumping on it, I could not straighten it out and was sent, escorted by a guard, to Obermassfeld Hospital near Meinigen by train, to have it operated upon.

This was a glorious diversion, despite the pain, dressed in best blue uniform for the journey, with a plentiful supply of cigarettes and chocolate for bribing, we entrained at Sagan station early one morning. I hobbled with a stick making the most of it!

Our train journey took us via Leipzig, where we had an hour or so to wait for the connection to Meinigen via Erfurt. Sitting with the guard in the middle of the centre platform in Leipzig station, I marvelled at the wonderful expanse of curved glass roofing above us and over the whole station. Not a crack or broken pane to be seen! Yet, according to a news item a day or so before, the RAF had bombed Leipzig station! Maybe we were on the wrong station?

After waiting about half an hour, a very long and dirty train steamed in to our platform and hundreds of German troops detrained, not with Russian snow on their boots but certainly covered in mud and filth. They rushed past us with never a glance, until one bright spark suddenly stopped and exclaimed, *'Ein Englander, ein Flieger'.* Several offered me cigarettes, some of their chocolate, others wanted to try their English. All were around twenty years old or so and they carried a heavy load of equipment, with the usual fur-covered haversacks. After a few minutes a couple of brown-uniformed officials with swastika armbands ran up and told them to disperse, with a few rude gestures and cursing they did so, leaving me in no doubt regarding their feelings towards these party bastards!

Soon we entrained for Meinigen. I was seated next to a green-uniformed forester and when lunch time came and I took out some rations to eat, with some Army-type dark brown bread and blood sausage – I was keeping the Red Cross goodies for later – he brushed them aside saying '*Nein, nein*' and insisted on giving me a hard-boiled egg, some white bread from a roll and some butter! War is a stupid business to be sure! I reciprocated later by handing round Players' cigarettes and Cadbury's best milk chocolate. Looking round the compartment, and indeed along the line, it seemed as though Germany was not yet fully geared up for the conflict compared to England. It was an incredible journey.

I had been asked to look out for anything of interest of course and watched the passing countryside carefully; it looked all too peaceful and rural with well kept fields, farms and villages. However, winding our way through a hilly region after Arnstadt, past Erfurt, I noticed several sidings with low loaders with some sort of weapon or such on them under canvas covers, then there were what looked like large doors with concrete supports, lights and walls in the side of the hill close to the railway. I made a mental note of the position.

Obermassfeld was an experience in itself, staffed by German and

British doctors and medical orderlies, the latter taken in France and at Dunkirk. It was almost packed with injured, wounded, British and American personnel, aircrew and others, and other nationalities of Allied forces. Group Captain Massey, the SBO, from Sagan was there at the time, his old wound from the First World War giving him trouble, together with injuries from his more recent forced landing when shot down.

I was soon operated on by an eminent London orthopaedic surgeon who did a wonderful job. Kept for a month or so to recuperate and get the leg back to normal I had a relaxed and englightening stay at Obermassfeld, meeting some of the bravest people ever. One in particular was an ex-trawler captain moblised for minesweeping, who had been blinded. He had been given a Braille typewriter by the Germans and taught Braille. He received many visitors and was an inspiration to the less disabled and younger men.

For the first time also, I met several American aircrew. It seemed that their daylight raids, as expected, were taking a heavy toll and the 'waist gunners' and the dorsal and ventral gunners too were the main casualties. Wearing their 'flak jackets', they were escaping body damage but receiving hits in the arms. These who had come down in France suffered most, as the French doctors who initially attended to them removed the damaged arms at the sockets instead of leaving a stump where possible. This annoyed and astounded the Germans, who considered it inhuman! ... and typical! Truly an enigmatic race.

Returning to Sagan, East Compound, from Obermassfeld I again spotted the hill containing what looked like an underground factory, or store, this time the canvas-covered waggons had gone but various lights were lit at the entrance. Arriving I reported this siting and pin-pointed the position. By this time the three Wooden Horse men had got out and all three had 'made it' and the escape was history to become world copy through Williams's book and the subsequent film. A magnificent effort indeed.

With the Christmas of 1943 approaching, there were many 'stills' at work and home-brewed raisin hooch being prepared. With the improved supplies of dried fruit the various syndicates, as allowed,

Author on left with Group Captain Ford-Kelsey, Squadron Leader Monaghan and Brazilian pilots. 226 OTU RAF Stradishall, 1953

The AOC-in-C Fighter Command, Air Marshal Sir Basil Embry, GCB, KBE, DSO, DFC, AFC saying 'goodbye' to the instructors, some staff and CO of 226 Meteor OCU RAF Stradishall, 1953.

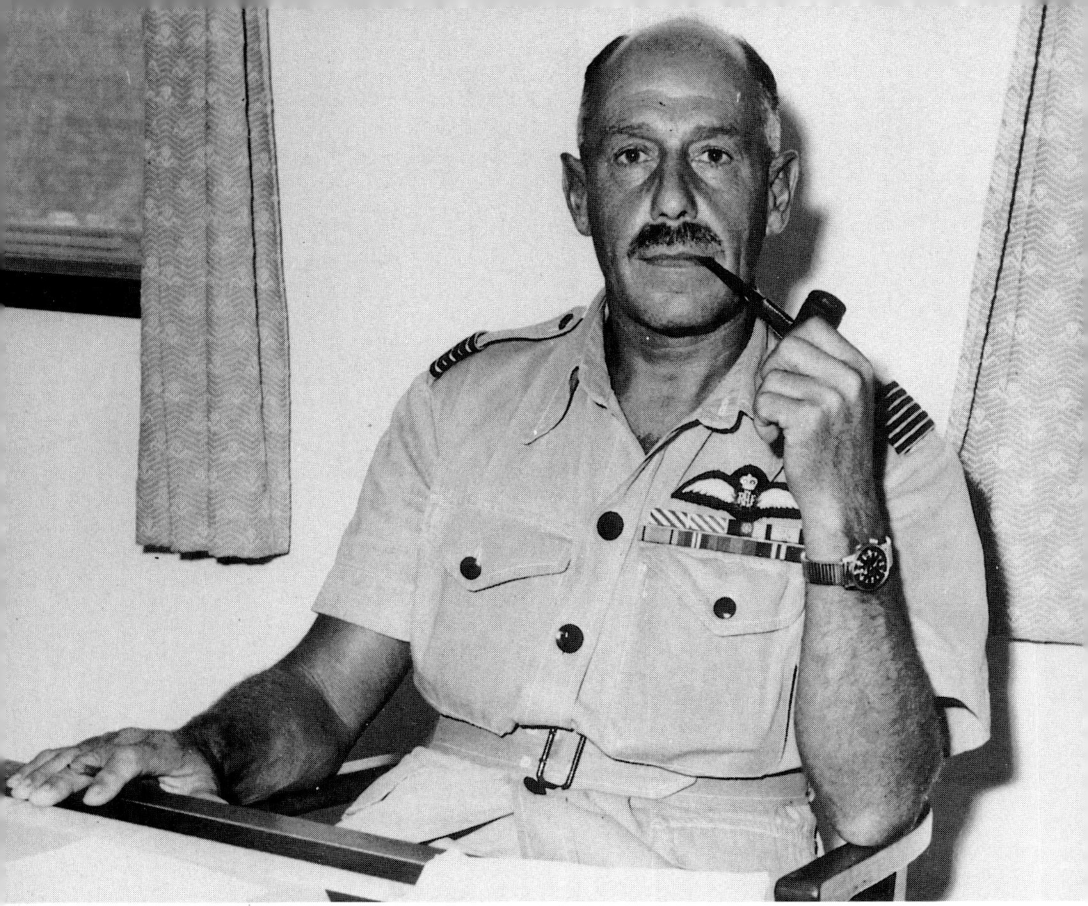

The author at Air Headquarters, RAF Persian Gulf, Bahrain. 1963–1964.

Middle East Headquarters, RAF Conference, 1955. Air Headquarters, Cyprus. Author first left, second row.

ıthor racing the MG 'A' at
etterton, 1958.

ıam Sprite' the first team of
stin Healey 'Sprites' raced
959. Left to right, Jim
ɔwnlee, 'Paddy' Gaston and
ɔ author at Goodwood.

ımbian Air Force De
villand 'Beaver' on the
ıks of the Zambezi.

Piper Navajo of Air Kenya Lt
at Victoria Airport, during its
construction, Mahe Island,
Seychelles, 1970. The first
ever scheduled air service
from Kenya to the Seychelle
operated for 2 years withou
incident. Discontinued whe
International Airport opene

Messerschmitt Bf109
retrieved from the English
Channel by the enthusiastic
members of the Brenzett
Battle of Britain Museum 3
years after it was shot dow
the author in 1940.

Another shot of the retrieve
Bf109 finally 'landed' near

were laying in stocks for the celebrations. This was going to be a vintage year, the news was getting better every day, morale was high despite the reduced tunnelling; at least in the East Compound. The theatre group was well ahead with its pantomime preparations. In the meantime, a new escape scheme materialised.

Within the regulations laid down in the Geneva Convention, private soldiers and junior NCO's were permitted to be used by the 'host' country in working parties, on railways, farms and such like. In the Sagan area there were several such parties and periodically men from these were brought into the sick quarters for treatment of some kind or another. The British Medical Officer was used to attend to them and 'visitors' from the main camp were allowed. The method was for a 'visitor' who was willing to relinquish his place in an officers' camp to find a 'look alike' who was willing to swap and take his place.

This took considerable organisation, exchange of relations, as it were and briefing on family background, special points and peculiarities. An exchange of identity for the German scrutiny and acceptance in the new environment by both men. A tricky business. However, one successful exchange was carried out by Flight Lieutenant McSweyn, an Australian who eventually made it back to the UK by this method.

Liaison with the North Compound was maintained and the 'X' and Intelligence Committees were getting better and better. The relations with the Germans remained on a good footing, the Kommandant, von Lindeiner; efficient, understanding and a good officer: Feldwebel Glemnitz, hard working, efficient, cheerful and uncorruptible. His head ferret, 'Rubberneck' was a different matter and heartily disliked, but good at his job.

Before Christmas, we heard that the new South Compound was going to be opened and all the American prisoners presently integrated with the RAF in the North Compound were to be moved into it. There were also rumours about another camp being prepared some twenty or so miles to the north. It was all go. The many 'X' 'factories' enjoyed better facilities than before and tame guards provided the best ever supply of contraband materials; information, identity cards for copying, maps, photographic necessities. The forgers, now unbelievably proficient, were turning out travel permits, documents, letters of introduction; you name it, they did it.

With a hard winter approaching, all activities slackened off and soon a blanket of snow covered the ground, restricting movement and inhibiting activities which required the camouflage of the dirty surface sand now hidden.

We went into a sort of hibernation; Christmas was celebrated with many boozy parties and the pantomime was a great success, though not quite as professional as when the Army provided the bulk of the actors!

Spring came quietly in 1944 and, together with a contingent from the North Compound, a large number of us were 'purged' to the camp we had heard about to the north of us, but in sight of Sagan, as we later learned. The transfer was brief and upon arrival we split up to select potential localities for the various necessary 'services', in relation to ease of surveillance, 'stooging' or closeness to the wire. The barrack blocks were reasonable, with good double-thickness walls in which to construct hiding places: partitions which could be adopted to provide 'banks' for tools, maps and all the contraband materials. I had become quite expert in constructing these by this time.

The 'box' arrived safely with its valuable cargo, so with others similarly equipped, we settled into Belaria well prepared. The contingent from the North Compound included several of the 'X' committee and principles of the team involved in the tunnel eventually named 'Harry', from which the ill-fated 'Great Escape' was made.

These naturally resented the move and arrived in a very bloody-minded state, as they had been deprived of their place on 'Harry's' roster; they felt that they had been given away. This was of course not so, just the luck of the draw or being known activists by the ferrets. A few months after settling in, we heard about the dastardly cold-blooded shooting of some fifty of the escapees out of eighty-three who had got out of 'Harry' in the North Compound in March. We learnt later that three reached England and the rest were returned to various camps. This decimated the 'X' teams and we were shortly ordered not to press the, now Gestapo and 'SS' dominated, authorities too far, the word was for discretion to be the better part of valour. We were told to concentrate on getting as many as possible repatriated, long term prisoners and others.

In the following months we made hiding places as required, including one specially large one under the theatre stage. This I made from scratch, purpose built, when the stage was being entirely rebuilt and the 'bank' could be designed into it. At the same time we received an enormous old type eight valve radio brought in under a load of manure for the gardens. This was reduced to its component parts and put away in the Theatre Bank in a matter of minutes along with the other goodies.

Another radio we had hidden in a chimney cavity in the sick quarters inside the camp. Kept at the bottom of the chimney and raised on guide wires when required, it sometimes had to be left up. Access to the sick bay was never queried so it was very convenient, though not strictly within the Geneva Convention.

One day we had a surprise 'blitz'; these entailed a block, or blocks, being surrounded by a posse of guards, evacuated and searched. If warned in time, we usually left some small bit of contraband to be found and once it was discovered they would be satisfied. There was an arrangement with a tame guard who would warn us after their morning briefing by leaving the block in the German Compound with his briefcase under his left arm if a blitz was planned, or under his right arm if not. He had no way of indicating which block, however.

On this day we had no warning, but as they formed up in their compound I had a feeling it was going to be the sick bay. I rushed over and jumped up through the trap door in the corridor ceiling, secured the radio to the bottom of its niche and left as they were approaching. The following morning I casually strolled over, ostensibly to visit someone, but really to check the trap door. When leaving, the German doctor stopped me and said, 'I'm sure they did not find anything important.' I was flabbergasted, not knowing how much he knew or was just guessing, but decided that someone else should do the checking for a while in case I was 'marked'.

From Belaria we actually saw our first jet aircraft, a Messerschmitt 262; we had previously heard this new and weird noise but never spotted what was making it. There it was, no propellers and obviously incredibly fast. It caused considerable speculation and conjecture and soon the sound became quite commonplace.

We also witnessed a deep penetration daylight raid by American

Fortresses when a widespread formation of twenty of them appeared well to the west of us, flying at about 18,000 feet. We watched the bursts of flak amongst them and then several being shot down, the fighters attacking them all but invisible. We heard later that they had been attacking a BMW engine factory some distance away, results unknown. It reminded me of our first escort mission to Cherbourg in 1941; it seemed a long, long time ago!

The war news was now getting better every day and we had regular news bulletins read out in the blocks almost every evening, everyone was 'tails up' and wondering how it would end. What was our course of action going to be, what instructions would we receive from the UK, what would the Germans do?

In October I was surprised, and delighted, to be told that I was included in the next batch for repatriation. This was the outcome of some collaboration with the MO, again our 'mad' doctor from Warburg and myself putting on an act, and it had worked.

We left soon after, meeting up with several batches from other camps: Army, Air Force, Navy, Allied Officers, old prisoners, wounded men and several sad cases of depression, mental problems and other symptoms. Several hundred in fact.

Arriving in Sweden in a matter of days, we were looked after almost too well; the food was too good and too much. After a few days, we embarked on the SS *Drotlinghome*, arrived in Liverpool on 10th October 1944, and were taken to a nearby Medical Centre where we were checked over, some interrogated, and sent on three weeks' leave. It was quite a homecoming.

CHAPTER FIFTEEN

Back to Flying

I took some of my leave to go over to Ireland and collect a spare 1939 Austin 8 that we had, see my parents and friends, get a game of rugby in (the knee now fully serviceable), and back to England with the Austin. The super old Lagonda had been sold whilst I was 'holidaying in Germany'. Back in England, I contacted some old chums at Headquarters Fighter Command, Bentley Priory, Stanmore and went to see them. I was welcomed back and told they could use me instructing on Spitfires immediately and that there were several on the Command Communications Squadron at Bovingdon and, 'You had better get some time in on them!' They rang the squadron and told them to expect me. Wonderful! Was my luck back with me again? I thought, as all those who came back with me and wished to fly again were put through refresher courses and God knows what.

I got in several hours' flying in the next week and was posted to Kirton-in-Lindsey, a Spitfire Operational Training Unit near Scunthorpe, with Hibaldstow as its satellite nearby. There I was made Flying Wing Adjutant whilst I got back into the swing of things. Obviously the RAF had taken on an entirely new look during the last three years. Was it three? It did not seem like it now; water under the bridge! New aircraft, new faces, new stations, expanded beyond all recognition. Not that I had really seen much outside Fighter Command previously, but the tempo was different, the mood serious, the equipment vastly improved and more of it. It was 'all systems go', with the end in sight.

The winter of 1944 was another hard one, particularly in Lincolnshire and flying was reduced in consequence. This was Number 53 OTU, equipped with Spitfire Mk V's, Master Mk2's and

Martinets. Their task was to convert newly trained pilots onto Spitfires and give them a basic grounding in operational flying. tactics and use of weapons. I was glad to see that the latter received greater emphasis than hitherto, with priority given to the ground support role for which the Spitfire Mk V was well suited.

The Master and the Martinet were new to me, the latter a modified Master converted and fitted with a target towing winch and used on the air to air firing range for towing a sleeve target. Both were delightful aircraft to fly, especially the former. Of all-wood construction, they now had the Bristol Mercury VIII or XV supercharged air-cooled radial engines, instead of the previous and lower powered in-line liquid cooled Kestrel. The rear, instructor's, seating had been modified so that it could be raised when required to give the instructor a better forward view. They were good aerobatic aircraft, but had to be 'flown' through all manoeuvres to ensure no high 'G' or aileron stresses were imposed and 'flick' manoeuvres were strictly forbidden. This version of the Master was in fact faster at low altitude than either the Hurricane Mk 2 or Spitfire Mk V. For this reason it was ideal for demonstrating fighter techniques, circuits and landings and gunnery to pilots from the SFTS's before and during their first experience on the Spitfire V's.

After three months there I was posted to the Central Gunnery School, at Catfoss, to become a qualified pilot attack instructor. This was a good course, on Spitfires, and I met many old Fighter Command friends. With intensive flying and some good rugby too – we won the Group Cup – I was getting back into the old environment and life was good.

Nearing the end of the course when flying 'pairs' with another fellow 'student' on our way to the air to ground ranges at Skipsea and having mild 'dog fighting' on the way, I was clobbered on my tail, or clobbered my partner, in hazy conditions. He forced landed with a miniature propeller, reduced to about seven feet in diameter, in a field with undercarriage down, saving the aircraft.

I was not so fortunate, or clever, as I had lost all control of my rudder and elevator control was restricted. However, I found that by maintaining a speed of about 120 knots I could keep fairly level, but any reduction or increase in power caused the aircraft to dive or climb and I could not control it. By using engine and aileron I reached the range and called the range safety officer and explained my predicament. I tried various power settings and tried to free the 'back end' but to no avail as I slowly flew round the range area. Meantime the duty NCO was watching from the ground and I was down to about 800 feet. I decided to bale out, and informed the range party. The run in area on the air to ground range was extensive, barren land and no harm would come to anyone.

Rounding towards the sea I set the throttle, jettisoned the hood and opened the side door and tried to get out, whilst trimmed as far as possible for straight and level flight. Standing up in the cockpit upset the trim, the parachute and dinghy pack knocked the control column, I was blown against the back of the cockpit. It was going to be difficult and I did not fancy the prospects of a wheels-up forced landing with so little control. Round again; this time I managed to get my left leg out onto the port wing, wangle the parachute and dinghy pack out over the side door space and get my right leg out as I held onto the cockpit side with my left hand.

The shift in weight and drag of my body on the port side made the aircraft wallow as I made a grab to close the throttle with my right hand as I pulled the rip cord, the parachute opened and I counted to five and hit the ground, landing heavily. Almost simultaneously the aircraft crashed in a crumpled heap only several hundred yards away; it did not burn. A lucky escape. Another 'life'?

After a cup of tea with the range party the ambulance arrived from Catfoss. It had in fact been summoned *before* I had baled out; the best service ever! This was one of the old pre-war Albion ambulances with the shiny brass radiator; sitting up front with the driver I set off back to base up the very narrow lane from the range. The high hedges on both sides obscured the view and at one sharp bend we went straight on through the hedge! Not my day? No damage and back onto the track. We got to Catfoss safely, where I was delighted to know that the other pilot had landed safely.

Some days later his aircraft was successfully flown out of the field after a new propeller had been fitted following a 'shock load' test on the engine. Mine was, regrettably, a write-off. Was this poetic justice after my tail ramming episode some three and a half years ago when downing the Bf109? Retribution perhaps?

Passing out as a pilot attack instructor, I returned to join the Gunnery Squadron at Hibaldstow, the satellite to Kirton in Lindsey. Using the Masters for dual instruction and Spitfires for 'pairs' dog fights, I really got back into the game and found that I had something of a flair for instructing, and also enjoyed it. The war was coming to a close and the need for replacements diminished so that we had a more relaxed time, but not without incident.

At Kirton and Hibaldstow I experienced for the first time the 'new' look Air Force with nearly as many girls, women, servicing the aircraft as men; it took some time to get used to them.

In very high wind conditions when operating the Spitfires it was necessary to find some means of keeping the tail on the ground when taxiing down wind. This was achieved by one or two ground crew sitting, or hanging on, the tail plane facing either the front or the back. Reaching the take-off point, the pilot would stop, wait to see the ground crew 'ballast' visibly off and clear before opening up to take off.

One morning a WAAF ground crew was bent over the tail facing backwards of a Spitfire Mark V; reaching the take-off point the pilot forgot about her, and despite the tailheaviness he should have experienced, opened up and took off with her tightly stuck to the leading edge of the tail plane by the airstream on her back!

Noticed immediately by all and sundry, and called by Flying Control, he did a quick, quiet, reasonably slow circuit and landed with her still in place. She nonchalantly got off, was picked up by the waiting ambulance and brought back to dispersal. There she seemed none the worse for the ordeal, smoked a cigarette, then it suddenly struck her. 'Oh, my God!' and fainted. This incident hit the headlines in the dailies the next day; apparently someone had alerted the *Daily Mirror* (who else) as soon as she landed. We never discovered who.

The commanding officer of the two stations was one Group Captain John Hawtry, who had spent considerable time overseas pre-war and had acquired a long brown bearskin coat with a difference; it had the fur side outside. In the winter, snow and rain, this coat gave off a very distinct odour. We liked the CO, but his coat during the winter months warned us of his approach from afar. Maybe it was just as well?

The news was good; May 8th and VE Day came with suitable celebrations on the station and in Scunthorpe. Number 53 OTU was closed and I was posted as a flight commander to the fighter OTU at Keevil, this time Number 61, near Trowbridge in a very pleasant part of Wiltshire. This time on the North American Mustang, built to a British specification and fitted with a Packard-built Rolls Royce engine, this was a superb aircraft. At the same time I had been given a permanent commission and was offered the opportunity to complete my engineering degree at RAF expense.

The prospect of a year of so studying after three years' 'extended leave' in Germany did not appeal to me, especially after getting back onto flying so quickly; furthermore I was shortly going to be married, so this offer I rejected, but accepted the first immediately.

Keevil was a wartime station, wooden huts, accommodation and offices, but with a sound background and great spirit. Flying the Mustang was a new experience; here was a long range fighter, purpose designed and built with a superb performance, good handling qualities and a sound gun platform. The Harvard Mk 2's which we used for dual instruction were noisy beasts, but fun to give dual in. Not as good as the Masters at Catfoss and Hibaldstow, I considered, but robust and easy to maintain.

Rugby was again good with Bath and Bristol nearby. I was soon playing for Trowbridge and enjoying it, the knee never looking back since the operation at Obermasfeld, though we made a great play on it for repatriation!

With peace-time upon us, we soon became car-conscious again and I did another swop; the Austin 8, with a little cash, for a superb 1932 Continental Two Litre Lagonda, much better than the previous one, all metal body, low chassis model. Tony Gaze, an Aussie instructor, had a streamlined Alta and Jimmy Broughton a lovely Alvis Silver Eagle two-seater. With the run of the engineering services we spent many satisfying weekends titivating these machines.

The Mustangs were unfortunately going to be replaced once again by Spitfires, so I decided to find out how long one of them could be flown with all the long-range tanks fitted. Carrying some food and a thermos in the roomy cockpit, which had arm rests and other amenities not associated with British fighters, I flew the length and breadth of England, Ireland, Scotland and Wales for some six hours or so. The engine never missed a beat, but I became quite exhausted sitting in one position for so long; a memorable flight.

We retained the Harvards for dual instruction and I began to really appreciate this aircraft because it demanded a high standard of discipline and awareness at all times that contributed to our training role immeasurably without us being truly aware of it.

We also had a rather lovely old De Havilland Rapide, or Dominie as it was called in the RAF, that glorious wood and fabric biplane with two Gipsy engines that purred. With the simplest instrument panel and controls, this was used as a communication aircraft and whenever a mess party was planned I took it 'up north' and collected a couple of crates of lobsters, cooked. Flying back we could not resist the temptation to crack a few claws and eat the flesh. This super old lady was replaced by a 'modern' Anson; well, at least you did not have to wind the undercarriage up anymore!

From Keevil I was selected for two very worthwhile courses; one to become a Qualified Flying Instructor, QFI, but at Flying Instructors' School, not the famous Central Flying School at Little Rissington, similar but on an abbreviated course and none the less demanding. This completed, a little later I joined the Empire Flying School course at Hullavington where the new concept of instrument flying and the instrument rating scheme was being taught, to produce instrument rating examiners. Here we had a wide selection of aircraft to fly, under some of the best instrument pilots in the Service, with the famous Prof Jimmy Kermode as the Chief Ground Instructor. Given the minimum, if any, dual instruction or conversion, we flew any of the following aircraft: Spitfire; Harvard 2; Meteor 4; Buckmaster; Lancaster; Oxford and Vampire in the most appalling weather, in fact searching for cumulo-nimbus to penetrate instead of studiously avoiding them! Most of those on the course were twin or multi-engine pilots with considerably more pure instrument flying experience than I, as a single-seater fighter pilot had, despite my tour on night fighters.

I frankly found the absolute dependence and reliance upon the attitude presentation by instrument interpretation difficult after our 'seat of the pants' flying in fighters. Although we had carried out many hours' cloud flying, and in my case night flying, in Hurricanes, our reactions, feel of the aircraft and approach to instrument flying were different to those from Bomber, Transport and Coastal, who had many hundreds of hours under these conditions.

It was an intensive course, but rewarding, with such a wide variety of aircraft and a broad ground syllabus. In fact flying took on a new dimension, with the ability to take off and land when even the birds were grounded! As the old adage goes, only birds and 'BF's fly and BF's fly at night and no one on instruments!

The course had one tragedy when Ken Driver, a South African who had also been a prisoner in Germany, was killed flying a Harvard. Our Chief Flying Instructor was also a South African, P. Retif, whose ability to fly in the worst possible weather imaginable had to be seen to be believed; a remarkable man. To fighter pilots, the use of the various landing and navigational aids was something of a revelation. Later, these of course have been improved upon, so that operating in conditions that once grounded us is commonplace, even to the extent of automatic landing, which was demonstrated to us in a basic form.

We passed out as instrument rating examiners and returned to our units to spread the 'gospel according to Saint Hullavington'! For a simple 'fighter boy', it was a hard sweat before I came to rely on what the instruments told me ... as long as they were working correctly. This time, for the first time, it had not come naturally. Was the luck running out? Not quite, but I felt it had stepped aside several times along the way lately.

Returning to Keevil, now commanding the Conversion and Tactical Squadron, I found that the Mustangs had been replaced by Spitfire V's and XIV's, the latter a lovely machine with the powerful Rolls Royce Griffon engine. We were also shortly to move lock stock and barrel to Chivenor, near Barnstaple in North Devon. Our last few months were not wasted; we flew hard and I also sampled the superb Spitfire XXI and XXII with the two stage Griffon engine and larger five bladed propeller, magnificent to fly and handle.

We were now designated Number 203 Advanced Flying School,

in reality still an Operational Training Unit. We moved in October and I flew a taxi service with the Anson every day from the 11th to 21st to and from Chivenor, which turned out to be the ideal location for this type of unit. Situated at almost sea level in Barnstaple Bay, with unobstructed approaches except for Braunton Hill just off the circuit, it had one of the, if not the, best weather factors in England. The air firing and bombing ranges were within a few miles and saved valuable flying time. The local area, being a concentrated holiday centre with sea, hotels, beaches and sporting facilities and friendly inhabitants provided a perfect background for high spirits and morale.

There I took over the gunnery squadron and was able, I felt, to impart some of the experience gained and the art of using the Spitfire as a weapon.

Whilst at Chivenor, we again had some very good rugby and won the Fighter Command Cup and some of us represented Fighter Command in the Athletics field. I attended several other operational type courses and in February 1948, was posted to the Air Staff at Headquarters, Fighter Command, and was proud to leave again with above average assessments.

Fighter Command Headquarters, Bentley Priory, Stanmore, with the renowned Air Marshal Sir Basil Embry, as Commander-in-Chief, was the experience of a lifetime. The Air Staff from the Senior Air Staff Officer, Air Vice Marshal David Atcherley, down to the most junior flight lieutenant was a galaxy of the most highly decorated, dedicated, energetic fighter pilots and navigators one could hope to gather together in one setting; each a character in his own right, day fighters and night fighters. What a team to join.

The Command Communications Squadron, where I had quickly got back onto flying on repatriation in 1944, was still at Bovingdon with a variety of aircraft at our disposal. We had the pick of the lot, except the C-in-C's Devon, for visiting units and stations and we tried to fly there in whatever aircraft the particular unit or station used.

With some food rationing still in force, I used to organise a source of supply of enormous whole legs of ham and crates of eggs from friends in Northern Ireland who were wholesalers. To collect these, we would arrange a weekend 'training and liaison' visit to

Aldergrove in an Oxford or Anson, bring them back and distribute the goodies; it was good flying practice and kept the spirits up. In those days we were encouraged to keep in practice and had the facilities to do so, both at Bovingdon and on the squadrons and training units, where we were welcomed and encouraged. It was still a time when any new type was a challenge and conversion usually only a quick run through the pilot's notes and a cockpit check. This would be frowned upon now and, in most cases, with good reason.

My posting to Fighter Command coincided with the expansion and re-organisation of the Command, necessary due to the deterioration of relations with Russia and the war in Korea. Our Commander-in-Chief, Sir Basil Embry, had a formidable task, but his dynamic personality, blunt and practical approach, backed by operational experience second-to-none, soon gathered an equally formidable team round him. His task was to expand and re-organise the Command which had been reduced to a skeleton of its former strength. This included the all-important Air Defence of Great Britain which, it was decided, required to be a total weapons system. Integrating the various separate components such as the early warning system, control and communications, fighter stations and squadrons, airborne radar and ground to air missiles into one entity.

I was responsible for the day to day running of the Day Fighter Operational Training Units; a minor role in this galaxy of talent, which included such names as: Ronnie Lees, Harold Maguire, Desmond Hughes, Derek Stapleton, Bob Braham, 'Paddy' Barthropp, Mike Giddings, Colin Coulthard and 'Jimmy' James, Denis Crowley Milling, John Ellacombe, Archie Potts (the Scientific Adviser), 'Buck' Courtney and the Command 'Flying Doctor' Doc Howitt, and George Westlake, who joined us later. All kept in flying practice and the competition for aircraft could be quite fierce sometimes.

The atmosphere was one of light-hearted co-operative efficiency, almost on a war-time footing, without the dangers. There was a serious dedication underlying the surface sense of fun which materialised in the number of the Air Staff who achieved Air rank and filled so many of the top-ranking appointments in the RAF in the years following. I shall never forget this period; it was a pleasure and an experience.

After a tour of two years, during which I was lucky to be selected for the Fighter Command rugby team, I was promoted and posted to Number 226 Operational Training Unit, at Stradishall, near Cambridge. There I took over as Officer Commanding the Flying Wing and Chief Instructor flying Meteor Mk 8's; 7's and PR Mk 9's with Martinets as target towers, the latter later replaced by Mosquito's suitably modified by Marshalls of Cambridge.

Alec Ford-Kelsey was the Station Commander and Colin Gray, a well known Battle of Britain pilot, was the Wing Commander Administration and Bill Allen the Chief Technical Officer. On the instructional side I had a great team of thirty instructors, most with operational experience, including target towing pilots, an American Exchange Officer and a Fleet Air Arm pilot.

Many of these were Sergeant or Master Pilots and one of our great disappointments whilst there was not being able to get more than one of them commissioned if they warranted it, and most of them did. The majority had war-time experience, several were well decorated and all were experienced, responsible pilots who had proved their worth, but according to the Air Secretaries Branch, and the Records Office, did not have the background or educational qualifications *when they joined*. This angered me because most of them were streets ahead of those we were training and, given the incentive, would have been even better, though older. The C-in-C supported us but Air Ministry was adamant. We lost some good men in consequence, as the conditions 'outside' were now becoming too attractive to resist. A situation which the Air Force and Treasury were slow to appreciate and redress.

At Stradishall, we produced about 30 fighter pilots per month, and two to four fighter reconnaissance pilots as required; these stayed an additional month. With a large establishment of Meteor Mk 8's, with the enlarged air intakes, and sufficient Mk 7's for dual instruction and conversion and gunnery training, we ran it like an Operational Fighter Wing. Flying, ground instruction, gunnery and sport at an intensive tempo averaging around 130 sorties per day, sometimes more. On short flights with a very congested circuit our accident record was unequalled, whilst the number of technical or engine failures was negligible. Personally, in one thousand Meteor hours I never suffered a major technical failure. A remarkable and under-praised aircraft.

Students arrived with eighty hours or so on Meteor Mk 4's from Flying Training Command, so the initial conversion was fairly simple. Converted and given tactical training at our Tactical Squadron, they graduated to the Gunnery Squadron, with one month at each. Each month we endeavoured to get the course leaving out a few days ahead of schedule to give us a couple of days for the Flying Wing to practise as a wing, as squadrons and have air to air firing competitions. We could in fact scramble 24 Meteors in two and a half minutes and match any operational wing in the Command.

Our sporting activities were extensive: all the usual games plus escape exercise, augmented by dining-in nights with the usual 'indoor sports' and any who ducked out copped it. We also trained several foreign and Commonwealth pilots from time to time, a highlight being a bunch of Israelis, a tough, likable lot, one of whom became their Air Force Commander-in-Chief. Determined, aggressive pilots with lots of guts, we even trained them to play rugby on the side! Others from certain countries were not so impressive; their attitude to discipline and efficiency appalled us, whilst their perfumed, foppish turnout made us wonder a bit!

At this time the efficacy of the National Service pilot training scheme was known to be under scrutiny and we were concerned at the waste of time and effort that 226 OTU was putting into training pilots who before, or soon after, the end of the course would leave the Service. The scheme had been introduced to provide a reserve of trained pilots and also a source of short service or long term pilots from those of them who liked the life and took to flying. In fact very few signed on and those who left had no means nor incentive to keep up their reserve status, whilst the introduction of faster and more sophisticated aircraft called for training time and conditions outside the scope of their National Service engagement.

This problem came to a head when we received an intake comprising 75 per cent of NSA pilots, nearly all of whom were being 'demobbed' at or before the end of their Meteor Operational Training and none were interested in signing on. I refused to train them and, taking the bull by the horns, bypassed our Group Headquarters and got straight onto Fighter Command Headquarters, knowing the C-in-C's feelings about the NSA scheme.

The outcome was that the C-in-C arranged a meeting at once at Stradishall attended by the then Under Secretary of State for Air, himself, our Group Air Officer Commanding, Air Vice Marshal Richard Atcherley (twin brother of the SASO at Fighter Command), our Station Commander and myself, together with the C-in-C's PA.

We stated our case, the C-in-C stated his, the US of S looked vaguely bored and did not seem to be listening, as he chewed his horn-rimmed spectacles. The C-in-C kept his cool admirably; I did not do so well and the AO C-in-C poured oil on troubled waters! The upshot was that our intake of NSA pilots was reduced and finally stopped, the Command benefited and the NSA scheme was doomed.

At this time, after the expansion of Fighter Command and the arrival of several American fighter squadrons, East Anglia was again virtually an enormous airfield, with crowded skies every day. We 'buzzed' them, they 'buzzed' us and met for dog fights with all the intensity of war-time. After several near misses we got together and called it a day before someone got killed. It was fun while it lasted, and mutually beneficial, but foolhardy.

During this tour I finally parted with the Lagonda Continental after it had done around 100,000 miles but was still in beautiful condition. This was followed by a series of mundane vehicles; a Morris Minor with side valve engine fitted with a Marshall blower which went like a bomb and surprised many at the traffic lights! A Rover 90, a nice, gentlemanly carriage which could, when pushed, give a good account of itself. A Jowett Javelin saloon with the boxer engine and torsion bar suspension, way ahead of its time, but doomed when their body manufacturers were bought out.

Paddy Barthropp was 'OC' Flying at Waterbeach, north of Stradishall and had custody of Hurricane Mk 2 Number LF-363 which he brought over for me to fly in 1952 ... but no violent maneouvres! Delightful, it is now one of the two remaining on the Battle of Britain Memorial Flight, 34 years later.

Towards the end of my tour, our Group Air Officer in Charge,

Air Vice Marshal Richard Atcherley carried out his last AOC's Inspection of the station and, after all the ceremonies, at the post-inspection wash-up paid me a great compliment, and slapped me down at the same time by saying, 'Mac, you are one of the best officers commanding a flying wing I have ever had, but you will never get on unless you are less blunt and more tolerant!' ... and I never did, either! At the end of my tour, however, I was awarded the Air Force Cross and again assessed as 'exceptional' as a pilot and instructor.

CHAPTER SIXTEEN

Get Your Knees Brown

From Stradishall I was posted to Headquarters Middle East Command at Ismailia, in the Canal Zone, Egypt, on the Air Staff as Wing Commander Operations and Training.

My arrival was somewhat spectacular, being met at Fayid by my predecessor, William Hoy, who flew me to Ismailia in a worn out Proctor, parked my kit in my room and took me to the bar just before lunch. Entering the bar I espied Don Gericke, last seen at Schubin POW Camp in 1944; he saw me and we joined in a rough and tumble on the floor, a right old wrestling match. Someone said, 'Oh, Christ' and several tried to separate us; we stopped our struggle long enough to tell them to 'Futsak', he being a South African. Honour declared even, we repaired to the bar for some well-earned beer!

Our Command territory from Ismailia extended to Cyprus, Iraq, Sharjah, Muscat and Oman, round the corner to Aden, including Salalah and Masirah and over to East Africa with the base at Eastleigh, Nairobi. We had detachments at various other places including training teams with local nationals and airfields at Amman and Mafraq in Jordan; there was great scope for keeping in flying practice visiting these places together with the RAF bases and squadrons in the Canal Zone. With the Canal Zone virtually under martial law and movement outside it curtailed, we made the best of the sporting facilities, which were quite extensive, available on our door step.

Once again, like Fighter Command, the Command, Group and Sector staffs in the Canal Zone were generally a selection of well decorated characters from top to bottom, as were the senior people on the many out stations. It was a no nonsense command with wide responsibilities covering a very large and varied area. I found it all exciting and interesting, never having been to the Middle East or Africa before, and took every opportunity to fly round our 'manor' so to speak.

The first visit, accompanying our Senior Air Staff Officer, Air Vice Marshal J.N.T. Stephenson, to Kenya, was a revelation. The Mau Mau rising was in full swing and we went to assess the operations undertaken from the RAF Station at Eastleigh, Nairobi, using Harvards as light bombers and Lancasters and Lincolns as heavies to smoke out the Mau Mau gangs hiding in the forests. Like looking for a needle in a hay stack the noise of the engines heralded their approach from afar in the silence of the Aberdare and Mount Kenya areas involved.

Upon arrival in the Operations Centre we were given a copy of the Mau Mau oath plus photographs of some of the atrocities perpitrated to examine to get us, as they said, 'into the right mood'. The AOC East Africa was Air Commodore Beisiegel, who escorted us round and looked after us very efficiently and introduced us to some of the most colourful characters imaginable; white hunters, farmers and police, several of whom I was to meet up with many years later when living there. They were worth a guinea a minute. It was a wonderful country with great potential, but ...

In November 1954, we had to evacuate the Canal Zone and hand it back to Egypt, so the headquarters was transferred to Nicosia Airport, Cyprus where, on a joint user airfield, we were well away from the civilian portion with a separate aircraft park for our communication flight and visiting aircraft.

As married accommodation had been limited in the canal zone for several years, many serving there had set up their families in Cyprus, where the RAF had a substantial presence at the airfield at Nicosia. The move was, therefore, welcomed by the majority, as it was really only a matter of expanding the existing facilities to Command level. Furthermore, those without their families there could now bring them; there was ample accommodation for renting and the environment was perfect for young children.

Initially, as on our freqent visits, life for all in Cyprus was idyllic, peaceful and friendly; we mixed freely with Greek and Turkish Cypriot alike. Kyrenia, with a large expatriate presence already welcomed the many families that chose to live there. With the excellent swimming, sailing and picnicking facilities it was akin to a permanent holiday situation.

I imported a Morris Minor Tourer as it was an ideal climate for an open car, with an uprated engine by Vic Derrington of Kingston, it gave a good account of itself. This was augmented by an MG Magnette, ZA, saloon – for £680 duty free – with which I won the road section of the 1955 Cyprus Rally. Even then this was a well organised two day event over some of the narrowest, twistiest mountain roads possible. My navigator was a retired Wing Commander, now a school master, with a computer-like mathematical brain – no 'Halda' required; all I had to do was drive, his averages and time-keeping being unbelievably accurate.

This idyllic set up soon vanished as the island erupted in turmoil, though the early stages of the Eoka campaign did not effect us too badly, except that we were carrying side arms again.

So, what's new? With terrorism condoned, accepted and supported by all the crack pots in the name of freedom or the right to express themselves. We carried on, but back to a strong security situation, and not knowing who was who. The rest is history and the problem still not resolved.

The Headquarters was renamed more appropriately the Near East Command and Cyprus got its own Group Headquarters with Air Commodore C. Boyce Commanding and Prosser Hanks his Group Captain Operations.

I left Cyprus and Headquarters Near East Command some months before the Suez crisis in March 1956. Before leaving I had the honour of flying the first RAF aircraft into the new airfield at Akrotiri, a Communications Squadron Pembroke from Nicosia, Cyprus, carrying members of the press in it. Not a very arduous flight!

Finally, just before leaving Cyprus I had a disagreement with our Senior Air Staff Officer. This was over one of our contingency plans for the area. I was glad to leave eventually, but naturally felt sure that this row had not helped my career prospects!

Returning to England I was posted to the, then Air Ministry, soon to become the Ministry of Defence (Air) and committed to a desk ... but not tied to it. This time in charge of the Operational Training in Bomber and Coastal Commands. Although an ex 'fighter boy' I soon got the hang of it all from my three squadron leaders and frequent visits to the Commands and their stations.

By now I had knocked up several thousand hours, on many types of aircraft without losing any more 'lives' and reckoned that I still had two or three to go. Even at the Ministry of Defence (Air) in 1956 we still had a Communications Squadron at White Waltham, near Reading, commanded by an old friend of mine, so I made good use of it. The most attractive aircraft they had to keep one's hand in on was the Balliol.

These had been introduced into the RAF as an Advanced Trainer, initially with various engines but with the Merlin finally, in small numbers, as it was in fact too late in concept and production to meet the Advanced Training requirements at that time. It was a pleasant aircraft to fly and aerobat and nice to get behind a Rolls Royce Merlin again and a relief from flying a desk. A noisy, rattly, but very robust beast with various peculiarities in the stalled condition depending upon the combination of power setting, undercarriage or flaps up or down, and air brakes in or out. The torque effect on 'gunning' full power at slow speed spelled trouble at low altitude, in short not an inspiring aircraft for new pilots and its early rejection as a trainer was justified. Nevertheless, lots of fun for us MOD (Air) chaps as long as it was treated with respect. I did not press my luck this time or even give the remaining 'lives' a tickle!

Back in England I renewed acquaintances in the motor sport world and decided to have a crack at motor racing, something I had always wanted to do. First spending some time round the various circuits, reading up the regulations and different classes of cars and events available, and renewing my several club memberships. Starting in club events was mandatory to meet the Royal Automobile Club Competition regulations, so I bought Watling-Greenwoods beautifully constructed 'RWG' with which he had won the prestigious Brooklands Memorial Trophy the previous year.

Hoping that the Irish luck and the remaining 'lives' would hold, I engaged in the usual beginner's round of the club circuits and races. Enjoying considerable success, I changed the car next year for an Elva Mk 2 with Coventry Climax 1100 cc engine. I was again reasonably successful until Colin Chapman's Lotus XI dominated the class and there were some changes in the regulations. When this happened I changed to an MG 'A' in the 'GT' class and received great encouragement and support from Marcus Chambers, the MG Competitions Manager; a great character, together with the 'Comps Departments' wonderful team of mechanics, still with the 'lives' intact.

It could not last, and in the Leinster Trophy Race in Ireland, on the Wicklow road circuit, the luck ran out ... or nearly did ... when leading after two hours. On a fast righthander, taken at about 85 plus mph, I hit a hidden culvert in the grass verge when clipping the corner with my right front wheel and rolled over. The car skidded upside down along the rough – and dirty – road surface, with me acting as a ball bearing between the car and the road! Coming to a stop – we did not have roll bars then – I crawled out from under with cuts, bruises, a dislocated shoulder and multiple abrasions, but alive. Another life gone??

If it had not been an MG 'A' I would not have survived; the strong, well-made chassis and frame saved my life and the luck of the Irish held. Marcus Chambers and his men were most helpful, the MG 'A' was repaired and back for racing in three weeks. I was repaired by the RAF at Uxbridge and back for racing at the same time and we went on to come third overall in the Autosport Trophy and win the 1500 cc class in the three hour final, after collecting points in the many qualifying rounds during the season, plus other successes in national events.

The following year we formed the first Austin Healey Sprite Team with great support from MG Comps Department again; Chris Tooley, one of their rally drivers, and 'Paddy' Gaston, ex RAF pilot and garage owner, joining me. We had a successful year, both individually and as a team.

I then changed to a Turner with BMC engine and was successful once more; the worst setback was losing the Dunboyne Trophy race in Ireland with a holed radiator in the last few laps. This car was followed by a Turner with a Cosworth Ford engine, a great car and very successful; my last race being the 1963 TT at Goodwood, finishing fourteenth, despite a broken half shaft halfway through several hours' racing.

During this period, whilst carrying out my duties at MOD, I became the first Chairman of the newly founded RAF Motor Sports Assocation. With my contacts at MG's, we acquired some Mini Coopers for the association, which they raced and rallied with minor successes and I raced the Turner under the RAF MSA's banner from time to time.

In this long tour at MOD I had moved onto the Organisation Staff after three years in Operational Training; this was an interesting time and a new departure for me, the task relieved by still keeping in flying practice at White Whaltham; a godsend and no more lives lost!

In November 1963, I was posted as Wing Commander Administration, Headquarters Royal Air Force Persian Gulf, in Bahrain, with Air Commodore Geoffrey Millington as commander. I knew Bahrain from the tour at Middle East Command and looked forward to this.

The evening I arrived at Muharrq, there was an Army cocktail party at their mess in Manama and I was invited along to meet people. The first introduction was, 'Meet the British Head of Customs'. We shook hands and he said, 'Well, Ken, long time since we were at school together.' Lionel Irvin, from Belfast. 'Welcome to Bahrain, now come and meet the Commandant of the Bahrain Police.' This time an older man ... 'Well, Kenneth, and how are your mother and father?' The ex-Head Constable Royal Ulster Constabulary, Enniskillen. So as they say, 'in like Flynn', the luck of the Irish holding good, lots of good Ulstermen around, even in Bahrain!

The Gulf was fascinating; the most unbelievable blending of the old and the new before the later oil boom and the frantic erection of the concrete jungles that now deface the shore-line. Bahrain, Qatar, Dubai, Abu Dhabi and Muscat still retained their old world charm, their own character and indefinable serenity that has gone now for ever. From Bahrain I was able to renew acquaintance with Beirut, Israel and Jordan, but going back is never the same; the media, speed of communications and politics had already begun to spoil the Middle East and make it an international cesspool.

In Bahrain we enjoyed a peaceful life with the Ruler and his people, being made more than welcome everywhere. There were no tourists, little development as such and life was exceptionally pleasant.

The RAF presence was considerable, on a joint user airfield on Muharraq Island – joined by a causeway to the main Island – with resident squadrons and many transit aircraft daily. There was also a strong military and naval presence with our headquarters at Jufair, just outside Manama the main town, not far from the naval jetty and the political adviser's residence.

Again, there were ample opportunities to fly and keep in practice using the Communications Squadron aircraft and borrowing squadron aircraft. Indeed, when I found the Pembroke with dual controls which we had had in the Canal Zone six years previously, it was like meeting an old friend.

Covering the Gulf, with a detachment in Kuwait, a large base at Sharjah and several airfields on the coast and round towards Aden, including the Oman, it was never a dull appointment for any of us. The Manama Sailing Club was very active and I was made Vice Commodore after a short time and bought a boat to use whilst there; with such outdoor life we soon became very fit and brown.

There were many amusing incidents, it was that kind of place, but generally none of them nasty. In the mess, which was on Muharraq Island, near the sea, we periodically held Scottish Dancing which was well attended by the RAF and civilians alike when invited. This was also attended by one or two of the Ruler's family, of whom there were many, and one of them, an elderly Sheik who was a regular attender, really surprised us one night. As PMC of the Mess I would meet them on arrival and in his case always ask how he was. He always replied, 'Oh, 50 – 50, some good some bad.' On the night mentioned, he arrived and I met him to find him in full Scottish regalia, kilt an' all, but under his own headgear. I pretended not to notice and asked the usual question, to receive a wonderful answer. 'Oh, 50 – 50, some good some bad, but from now on you must call me Sheik *Mac* Rashid!'

They were lovely people really, compared to some others in the area, and our visits to the various Gulf States were always pleasant and interesting, if hot; never boring and always welcome. The tour was overshadowed by the knowledge that the UK Labour Government, in their churlish 'wisdom', or to please or appease you know who, were thinking of withdrawing the British presence from the Gulf – as they did, with disastrous results.

When the decision was told to the Gulf States it caused great concern amongst them, because as long as Britain policed the area, in their opinion there would be peace and stability, and where these exist there is trade and prosperity, so why leave? The Ruler predicted chaos and also feared Iran, who had ambitions towards Bahrain from 'way back'. They went so far as to suggest that we stay and they would subscribe, or even pay for, our presence. Naturally, this was rejected and the Ruler's predictions materialised, later than he predicted maybe, and the stability of the whole region has deteriorated and the rise of Islamic Fundamentalism given a free reign.

Whilst in Bahrain I was fortunate enough to undertake two 'clean up' operations for the AOC-in-C, Johnny Johnson. The first as OC Masirah Island, which was an experience in itself as far as the local inhabitants were concerned. With two factions of Arabs, the Sheik on one side of the island and the Wali on the other, I was frequently called upon to arbitrate between them on local disputes; it was like a piece out of the Arabian Nights. They, and their people, were poor as church mice but had such tradition, character and customs; lovely people.

The second was as OC RAF Eastleigh at Nairobi, in Kenya. There, I renewed my acquaintance with the RAF, who were later to leave, and the Kenya Air Force, still RAF staffed and controlled, but also soon to become autonomous with a small RAF presence as advisors only when Eastleigh was handed over to them.

During my tour at Eastleigh, there was considerable operational activity in the remote and arid areas on the Somali and Ethiopian borders against the 'Shifta' gangs, where we were supporting the Kenya Army, and some police. This gave me the opportunity to do quite a lot of flying in the De Havilland Beavers and Scottish Aviation Twin Pioneers, both STOL aircraft and great to fly on reconnaissance, supply and casualty evacuation flights.

Kenya was a wonderful country and, with no more 'lives lost', I resolved to go back sometime; little did I know what was ahead.

From Eastleigh, I was posted to Headquarters Maintenance Command at Andover, despite Johnnie Johnson promising me a good posting and promotion for the good work we'd done for him. However, as he, my boss in Bahrain and I all retired from the RAF at about the same time, I do not think that we quite fitted into the Air Secretary's 'new look' requirements anyway. We did not conform!

At Andover, the only saving grace was having the Command Communications Flight on the airfield next door to the Headquarters, without which this posting on the Organisation Staff would have been unbearable. In fact it was relieved when Ian Smith declared UDI in Rhodesia and I was sent with the RAF detachment to Lusaka, in Zambia, with Jimmy James as our AOC detachment, an old Fighter Command and rugby 'mate' of mine and a great character and flyer. This short tour would fill a book of its own. My luck holding good, we had the use of the Zambian Air Force, RAF staffed and run, De Havilland Beavers.

Using a squadron of Javelins and Mobile Radar, we were supposed to watch the Rhodesian Air Force whilst our Transport Command Britannias ferried oil in from Nairobi. There was never a dull moment. Our headquarters was in a rectangular single storey building which had been a government hostel, opposite the Lusaka Club, very conveniently right on the approach to the 'old' Lusaka airfield. With only one runway, and this a short one, the 'Brits' did a very efficient job and the standard of flying, schedule keeping and serviceability of these aircraft and crews deserved more credit than they ever received.

Whilst there, the then RAF Commander of the Zambian Air Force and their Defence Secretary asked me if I would consider taking over as commander, as the President wanted to have 'his own' Air Force and they would shortly be recruiting to this end. This meant retiring from the RAF prematurely which, at this stage I was not averse to doing, as I felt that I had reached a dead end. Furthermore, as Org 1 at Maintenance Command I had had my first brush with the unions and civilians manning sixty per cent of the units in the command. What I had seen and been introduced to in their manning, working hours, restrictive practices and inefficiences, not only in the civilian establishment either, had blunted my already waning enthusiasm and respect for the Service, and Government, I was serving and had loved.

CHAPTER SEVENTEEN

African Finale

After several months, whilst my retirement was going through, the Zambian High Commission said that the President now wanted a 'retired Air Officer' to match the rank of their Army Commander, and could I find one? As luck would have it, my ex-boss in Bahrain was also retiring, so I put the Zambian Defence Attache (who was an ex British Army Officer), in touch with him and eventually we both arrived in Lusaka. He became the Commander and I his Chief of Staff and Deputy Commander with a staff of retired RAF officers and NCOs and some RAF regulars still attached to the ZAF, who heartily resented our arrival and did little to make us welcome; not that it worried us, only discredited them.

In the event, our tour of duty was an experience, to put it mildly; we had many problems never previously encountered involving a different culture, way of life and even tribalism. The ZAF was then equipped with De Havilland Caribou and Beaver aircraft in the Transport Role and the same maker's Chipmunks for elementary flying training, augmented by some Dakota's and a British Aero Space HS 748 as the President's aircraft. It was another world to us in many respects and we were not sorry when there was a disagreement between the President and HMG and he decided to get rid of the British presence and told us to recruit ex Indian Air Force personnel to take our places, which we did in 1970.

Simultaneously, the Italian Air Force was asked to take over and run the Flying School at Livingstone, doubtless influenced by the fact that Al Italia ran the Zambian Air Line at that time. Upon arrival they insisted on having separate accommodation, messing and refused to 'fraternise', not that we wished to do more than was essential for some degree of efficiency with people whose motives were suspect and appearance varied from the filthy dirty to the absolute ponce. They eventually spread to Lusaka when the ZAF was 'persuaded' to purchase Agusta Bell Helicopters; where was Westlands of Yeovil? We, naturally, endeavoured to push British aircraft and components but were unimpressed by the lack of drive or enthusiasm by the British Aviation Industry compared to their continental competitors to the detriment of UK exports.

Nevertheless we did our best to produce a reasonably efficient and disciplined air force, but the conditions, the political climate and nuances together with financial and local problems were against us.

At the end of our three year stint we handed over to a heterogeneous mixture of African, Indian and Italian Air Force personnel and wished them the best of luck. God knows they needed it. We departed without regret.

During our tour we were all subject to nightly burglars, raiders to steal wheels and tyres off parked cars and such like. On one memorable occasion one of the newly arrived replacement officers, who were allowed to carry their side arms, heard someone at his car outside his house. Getting out of bed he looked out and espied some men round his car, so in his pyjamas he grabbed his little Beretta and opened the front door and was about to shoot at them, or over their heads, when his pyjama trousers slipped down! Reaching to recover them and cover his embarrassment – or something – he shot himself in the thigh. We thought that there must be a moral in this somewhere?

Apart from the ZAF I enjoyed some good motor rallying while there; one rally in particular, the Shell 1500, in which I had a glorious run in an 1100 cc 'sports saloon' Subaru with torsion bar suspension and a fabulous light alloy 'boxer' engine, entered by the local agent. Unfortunately, when very well placed, being side-lined with a broken wishbone in the front suspension at the penultimate checkpoint.

During this tour in Zambia, I invested in some property in Kenya and when leaving the Zambian Air Force, with no more 'lives' lost, was asked to take over as managing director of a large civilian light aircraft charter company. This had bases at Nairobi, Malindi, Mombassa and Nanyuki. The aircraft were a mixture of Piper, Cessna, Beech and several hired during the peak periods, including an Islander. The pilots were a mixture of local ex-patriots and ex-RAF. The local chaps being Kenyan born, had a wealth of 'bush flying' experience and knew the country backwards. Our main task was flying tourists round the game parks, to and from Uganda and Tanzania and commercial charter work, the operating being in and out of the small and rudimentary airstrips with which East Africa abounds. Dead Reckoning (DR) navigation and local knowledge were the order of the day, as navigational aids were minimal and approach or landing aids nil, except at the few International Airports.

The standard of flying and navigation was of necessity good and decision making the prerogative of the captain. Weather changes in the rainy seasons were rapid and unpredictable and could be dangerous in the vicinity of high ground, whilst the fact that Nairobi was at 5,800 feet or so to begin with, meant that the upper margin before oxygen was necessary was limited. For this reason, climbing above the weather was usually impossible and even if one had been able to do so the paucity of bad weather aids would have made it hazardous to get down again. Therefore, despite the apparent 'gin clear' flying conditions so often quoted by visitors, flying in East Africa was not a bed of roses and 'bush pilots' not the rather hairy types that many make them out to be, but in the main a very efficient bunch of chaps – and girls, for we had several very good lady commercial pilots there.

Having said this, the appalling accident rate in flying generally in East Africa cannot be overlooked. This was due to the inefficiency of the Civil Aviation Authorities and poor supervision of a small minority of operators and pilots, both military and civil, which at times overshadowed the excellent performance by the majority of very professional companies.

This appointment was a new challenge and enjoyable to start with, until the 'bag of worms' emerged; the varied interests of different shareholders, including the Government who were major participants, and internal squabbles. Wilson Airport, where we were based, was the home of all the light civilian aviation in Kenya, with the main agencies for all the major manufacturers overseas. They provided servicing, complete overhaul facilities, spares and training facilities. Mainly staffed by ex-patriates in every facet and function, as there were just no suitably qualified Africans available, they were the targets of constant attack for 'non-africanisation'.

This caused further friction which we could well have done without and aviation there suffered because of it. Wilson Airport itself was a hive of intrigue, with too many facilities chasing too few customers most of the time; however they all seemed to survive. The commercial world was new to me and I had to learn the practices, and mal-practices, quickly. As soon as possible, we obtained the Beechcraft Aircraft Agency and I frequently visited their factory at Witchita. Beautifully designed and built aircraft they were; they are in my opinion the 'Rolls Royce' of light aircraft. However, with a Board of Accountants and Travel Agents, we were never able to exploit this agency as planned.

One noteworthy project was our initiating the first ever scheduled service to the Seychelles, whilst there was a British Governor there. Using Piper Navajo's, we flew from Mombassa to Diego Suarez, night stopped there, and flew on to Mahe the next day. For this flight we carried two pilots and fitted the aircraft with ex-RAF 'Drift Sights', as we had no other way of getting 'winds' over the sea. On Mahe Island we landed on a small strip marked out on the land being reclaimed for the international airport in Victoria Bay whilst the new airport was being constructed.

After operating for several months, we got permission to use the small airstrip on Astove Island, a tiny island in the Aldabra Group, where we installed a beacon and Southern Line Shipping from Mombassa stocked aviation fuel and oil there for us. By operating through Astove we cut out the Diego night stop and could make the Seychelles in one day.

Following the success of this operation, we were asked by the Seychelles to look into the provision of an Inter-Island Air Service. For this we took a team there and did a survey of the many little islands which had marvellous tourist attractions, found those suitable for using and prepared a very complete report and proposal, with schedules, costings, etc., to commence using Britain Norman Islanders. The scheme was rejected by our Board as it conflicted with certain of their tourist interests in the Seychelles; passed to a rival company at Wilson Airport it became a successful operation in no time.

At this time I received several requests to represent various overseas companies who had interests in East Africa, so I set up as an Aviation Consultant. After a few more flights for pleasure, with the 'Irish luck' holding good and maybe one more 'life' left, I decided to call it a day. Over 4,000 flying hours on some sixty different types of aircraft; eight, or is it nine, 'lives' lost, but not one engine or major technical failure throughout, in war and peace and all kinds of weathers. Visiting many countries, meeting many wonderful people and flying many and varied aircraft, some good some bad, especially the superb Hurricane in Combat. In this respect let no one be deluded by the media and the publicity attributing the success in the Battle of Britain to the Spitfire. Good as the Spitfire was there were two Hurricanes to every Spitfire and they took the brunt of the battle. The Hurricane contribution was proportionally even greater as they were more rugged and repairable after crash landings or battle damage. An exciting life, no regrets and none the worse for it – with one 'life' remaining! Per Ardua ad Astra.

Glossary of Fighter Terms

Angles:	Height in thousands of feet.
Balbo:	Full squadron effort.
Bandit:	Enemy aircraft.
Buster:	Maximum cruising speed.
Gate:	Maximum possible speed, at maximum boost, permitted for 5 minutes.
Gyro Sight:	Gyroscopic Gun Sight.
Readiness:	State of Operational Readiness ordered by Group, or Sector for Squadrons to be able to get airborne in. For example, 'Standby'; 15 Minutes; 30 Minutes, and sometimes 'Cockpit' readiness with the pilot sitting in their aircraft ready to start up and take off.
Scramble:	Get airborne as quickly as possible.
Stood Down:	Not on call for Operations.
Pancake:	Return to base, refuel and reload.
RT:	Radio Telecommunications.
IFF:	'Identification, Friend or Foe'. Own aircraft 'blip' discernible on the Radar Screen.
Controller:	Fighter Controller directing the interception.
Cannons:	20mm. Cannon, Oerlikon or Hispano.
Machine Guns:	.303 inch. Browning machine guns.
Weaver:	Before the introduction of 'Pairs' or 'Finger Four' Operational formations 'Weavers' protected the 'blind spot' to the rear of Squadron formations against surprise attack.
Green Stuff:	100 Octane aviation fuel, coloured green for identification.
Trolly Acc:	Wooden trolly with starter batteries and leads for starting the engines.

Appendix A

For the Record

1935	North of Ireland Aero Club		Newtonards	Pilot
1939	No 24 E&RFTS (RAF)		Belfast	Sgt Pilot
28.12.39	No 3 ITW		Hastings	Sgt Pilot
25.2.40	No 5 EFTS		Hanworth	Sgt Pilot
25.5.40	No 3 SFTS		South Cerney	Officer Cadet
30.8.40	No 6 OTU		Sutton Bridge	Pilot Officer
21.9.40	No 43 Squadron		Usworth	Squadron Pilot
27.9.40	No 501 Squadron		Kenley	Squadron Pilot
4.10.40	Ju88 Destroyed	½		
5.10.40	Bf109 E Destroyed	1		
7.10.40	Bf109 E Destroyed	½	Same combat	
7.10.40	Bf109 E Destroyed	1		
25.10.40	Bf109 E Destroyed	1	Same combat	
25.10.40	Bf109 E Destroyed	½		
25.10.40	Bf109 E Damaged			
27.10.40	Bf109 E Destroyed	1		
29.10.40	Bf109 E Destroyed	1		
30.10.40	Bf109 E Destroyed	1		
9.11.40	Bf109 E Damaged			
14.11.40	Bf109 E Damaged			
15.11.40	Bf109 E Destroyed	1	Same combat	
15.11.40	Bf109 E Damaged			
23.11.40	Ju88 Destroyed	½		
28.11.40	Bf109 E Destroyed	1		
27.12.40	No 501 Squadron		Filton, Bristol	Deputy Flight Cmdr
8.4.41	No 501 Squadron		Colerne, Bath	
19.6.41	No 247 Squadron (Night Fighters)		Predannack	Flight Cmdr
7.7.41	Ju88 Destroyed	1		
13.9.41	He111 Destroyed	1		
31.9.41	POW Germany			
10.9.44	Repatriated			
19.12.44	No 53 OTU	(Instructor)	Kirton in Lindsey	
17.6.45	No 61 OTU	(Instructor)	Keevil	Flight Cmdr
8.2.46	No 203 AFS	(Instructor)	Keevil	Squadron Cmdr
16.10.47	No 203 AFS	(Instructor)	Chivenor	Squadron Cmdr
1.2.49	Headquarters Fighter Command		Stanmore	Air Staff
13.7.51	No 226 OCU	(Chief Instructor)	Stradishall	Wing Cmdr

28.8.53	Headquarters Middle East Air Force	Ismailia Nicosia	Air Staff
9,4,56	Air Ministry	London	Air Staff
27.11.63	Headquarters, Persian Gulf	Bahrain	Admin. Staff
27.7.65	Headquarters Maintenance Command	Andover	
28.6.65	Headquarters, COMBRITZAM	Lusaka, Zambia	
29.,9.65	Headquarters Maintenance Command	Andover	
30.6.67	Retired RAF at own request		
14.7.67	Deputy Commander Zambian Air Force	Lusaka, Zambia	
31.4.70	Retired Zambian Air Force		
8.6.70	Managing Director 'Air Kenya Ltd'	Nairobi, Kenya	
1.4.73	Retired Air Kenya		

Royal Air force Courses Completed
Pilot Attack Instructors, Catfoss
Flying Instructors School, Lulsgate Bottom
Empire Flying School, Hullavington
Officers Administration School, Digby
Offensive Support, Old Sarum
Air Defence, Old Sarum
Transport Support, Old Sarum
Amphibious Warfare, Fremington

Victories

Destroyed Day	8
Destroyed (Halved)	4
Destroyed Night	2
Damaged	4

Qualifications RAF
QFI
PAI
IRE

Qualifications Civilian
Instructors endorsements singles/multi's

Aircraft Flown (in Alphabetical Order)
Avro Cadet
Avro Tutor
Anson 1 to 30
Audax
Aztec.
Balliol
Beagle Twin
Beagle Pup
Beaver
Buckmaster
Beech Baron 55 & 58
Beech Bonanza
Beech King Air
Beverley 2nd Pilot
Caribou
Cherokee 6/260; 6/300; 140
Commanche 180; 250
Commanche Twin
Devon
Dominie
Dakota
Fox Muth
Harvard
Hurricane 1; 2; 'B'; 'C'
Hart
Hind

Hs 748	Proctor
Hunter 6	Prentice
Islander	Percival Q6
Lancaster	Pioneer Twin
	Pembroke
Magister	
Master 1, 2 & 3	Spitfire 1, 2, 5, 9, 14, 16, 21, 22
Martinet	Seneca
Mustang	Skyvan
Monospar	Skyservant
Mosquito 6	
Meteor 4, 7, 8, 14	Vampire
Moth	Venom
Navajo	
Oxford	

Operational
290 Operational hours, over 300 sorties (Inc. Training) Sept 1940 to Sept 1941

Assessments

1938–39	as a pilot – Above Average
1940	as a pilot – Above Average
1948	as a pilot and Instructor – Above Average
1950	as a pilot and Instructor – Exceptional
1953	as a pilot and Instructor – Exceptional

The following served with 501 Squadron at one time or another during the Battle of Britain. Of these some 45 were either killed, wounded or posted for various reasons. Many more were killed later in the war.

Adams H.C.*	Fenemore S.A.*	Holroyd W.B.
Aldridge K.R.	Gibson J.A.A.	Houghton O.V.*
Barry N.J.M*	Goth V.*	Hulbert D.J.
Bland J.W.*	Glowacki A.	Jones D.A.E.
Cox P.A.N.*	Green W.J.*	Jarrett R.W.E.*
Crabtree D.B.*	Gent R.J.K	Konrad.
Dafforn R.C.	Hairs P.R.	Kozlowski F.
Dixon F.J.P.*	Harrold F.C.*	Laws G.G.S.
Don R.S.*	Hewett D.A.*	Lee K.N.T
Duckenfield B.L.	Henn W.B.	Lacey J.H.
Egan E.J.	Hogan H.A.V.	Lennard P.L.*
Edge E.J.*	Holden E.	Loverseed J.E.
Ekins V.H.	Howarth E.F.	Lukaszewicz K.*
Farnes P.	Holland A.L.	Lonsdale R.V.H.

McKay D.N.E.	Pearson G.W.*	Sylvester E.J.H.*
Mackenzie K.W.	Putt A.R.	Stoney G.E.B.*
Murray A.D.	Patterson G.N.*	Vlad
Michail	Rogers E.B.	Whitehouse S.A.H.
Muchowski F.	Snell V.	Wilkinson K.A.
Morello F.	Saville	Witorznec
O'Byrne P.	Saward C.J.	Zenker P.*
Parkin E.G.	Skalski S.	Rose-Price A.T.*
Pickering T.G.	Storrat	Van Den Hove.*

* known to have been killed in the Battle.

Appendix B

Combat Reports

Author's Combat Report for 7th October 1940. Two Me109s were destroyed (one shared)

When on patrol with No 501 Squadron I followed Red 1 in attacking an Me109. Red 1 emptied his guns and damaged the enemy aircraft hitting his glycol tank. Red 1 (S/L Hogan) broke away to the right as the enemy aircraft entered low cloud at 4,000 feet. I flew above Red 1 below the bank and the E/A came out above at right angles at about 5,000 feet. I was above him at 6,000 feet to 6,500 feet and attacked from astern. He was losing height fast and went into the sea off Hythe from 3,000 feet. The E/A sank slowly the tail being visible for some time. I circled twice and saw the machine sink and then flew back to the coast climbing to 23,000 feet. On reaching this altitude I could not see any sign of my squadron but heard position of enemy aircraft over R/T and decided to patrol between Dover and Folkestone at this height, about 8 to 10 miles inland. At about 1335 hours I saw 8 Me109's coming across the coast from the East about 1,800 feet above me. I attacked the three rearmost machines in vic formation from beneath and a fourth enemy aicraft doing rear-guard flew across the line of fire and he developed a leak in the glycol tank. He half-rolled and dived towards the coast. I followed him, and his aircraft was only about 200 yards and so was easy to catch. I emptied the rest of my ammunition into him from 200 yards but he still flew on and came down to 80, to 100 feet off the sea. I flew around him and signalled him to go down, which had no result. I therefore attempted to ram his tail with my undercarriage but it reduced my speed too low to hit him. So flying along side I dipped my starboard wing-tip onto his port tail plane. The tailplane came off and I lost the tip of my starboard wing. The enemy aircraft spun into the sea and partially sank. I did not wait to see the final plunge as two Me109's were attacking me from above and behind. I flew back to Folkestone at 20 to 30 feet above sea-level and weaving up and down. They emptied a lot of machine-gun and cannon fire into my machine and damaged the engine, which belched oil and smoke into the cockpit. They broke off the engagement about one mile from the coast. I managed to make the coast and forced-landed between Folkestone and Hawkinge. My aircraft was pretty badly damaged and shot up.

Author's combat report for 13th September 1941. One He111 was destroyed

I was ordered to patrol convoy W. of Pentreath, but was changed to patrol base, then vectored 300 at 5000′ orbitted point 12 and ordered to 5000′ saw AA over convoy over which was 10/10 bank of cloud around 3000′ – 4000′. I investigated and on approaching cloud bank saw one A/C emerge from cloud going S. at 3000′. I turned and followed, closing and identifying as He111 at 1 mile he reduced height to 800′ and I closed to 600 yds and opened fire and missed, return fire was experienced but was high and what looked like 2 Very lights orange. I closed to 300 yds and fired up fuselage and starboard engine which cut and E/A swerved to right, was righted and I gave burst into port engine and E/A side slipped into sea, portwing entering first, pieces came off and hit my A/C tearing small hole in fuselage fabric. I circled and watched tail slowly sink and a piece of the E/A appeared to be floating. Cannons of IIC very effective indeed.

Squadron Offensive Patrol Report for 29th September 1941, when the author was shot down.

4 a/c Whirlwinds II 263 Squadron, escorted by 9 Spitfires IIs. 313 (Czech) Squadron, took off Predannack 1830 to attack Lannion aerodrome. Crossed English coast 1837 and crossed Channel on course 150 degrees at sea level, making landfall at Sept Iles 1905

hours. Followed coast to east of Ile Grande and turned SE to Aval Bay direct towards Lannion aerodrome approaching from NW. The Spitfires left the Whirlwinds 3 miles NW of the aerodrome and circled.

F/O Coghlan led the attack in a right-hand circuit diving down from 500 feet giving one 2-secs. burst at north dispersal point, and one 2-secs. burst at a Ju88 parked on S side of runway on SE end. Hits seen and a/c seen on fire claimed destroyed. Made a second circuit finishing up flying NW north of runway giving a final burst to hangars in north dispersal point, then returned N and W recrossing coast at 1912 hours. Made landfall near Mansolo Pt. and turned towards Predannack. Both engines cut when approaching to land and force-landed 500 yds E of aerodrome at 2030 hours.

A/C Category III. Pilot Unhurt.

P/O Warnes followed in to attack diving down from 500 feet firing at N dispersal point. Continued in right-hand circuit firing at S dispersal area where he could dimly distinguish a/c in dispersal pens. Poor visibility and long shadows prevented any attempt at recognition or estimate of numbers. Went out to W of aerodrome then N, re-crossing coast near Ile Grande, made landfall at Lizard and landed Predannack 2005 hours. Both pilots report heavy return fire from the ground, chiefly M/G fire from the ground all round the aerodrome, some black bursts above. (query explosive cannon shells). F/O Coghlan reports intensive heavy flak from wood NE of aerodrome.

Sgt. Maddocks made approach to target, turned to left, lost his No 1 saw shower of A/A (puffs and tracer) but couldn't definitely see the aerodrome and came out without firing his guns, returning home rather more to the west. (Is not sure of place where he crossed the coast), and landed Portreath 2005. Sgt. Hunter was heard on the R/T to say he was going to bale out, having made land near Plymouth, but no reports of him have been received from Observer Corps or police. No o/a seen in the air by the Whirlwinds. The weather was hazy. 10/10ths cloud over aerodrome (height of cloud not known). F/O Coghlan and P/O Warnes report escort satisfactory.

Red 1 of 313 Squadron escorted the Whirlwinds over the target and saw attack by 3 Whirlwinds – results not observed. The Spitfires escorted 2 Whirlwinds on the return and landed Portreath 2010 hours. 4 Hurricanes IIc 247 Squadron took off Predannack 1850 hours to follow the Whirlwinds. Channel crossed at 200 feet course 152 degrees. Crossed French coast at 1915 hours just east of Les Triege Islands. Went in over Lannion aerodrome roughly in line abreast at 100 feet. Red 1 (F/Lt Smallwood) and Blue 1 (*F/O Mackenzie*) slightly ahead. Red 1 selected the perimeter track on west side and fired a long burst at what appeared to be dispersal bays with no visible result. Flak opened up as Red 1 fired and was so intense and accurate that observation was difficult. It came from all sides of the aerodrome. Red 1 left the aerodrome at SW corner and looking back saw an explosion and a large crimson flash with dark smoke on the ground on NE side of runway. Set course for home encountering flak for considerable time after leaving target. We left coast due W of Lannion and landed 2015 hours. Red 2 (F/O Hordern) went in behind and to right, fired at AA post on S side and at a second one further W silencing the second with direct hits. At this stage Red II's a/c was hit in right wing and pilot slightly wounded so he turned away across Lannion town and down over the river at 600 feet being followed by flak the whole way. Turned N and left French coast over Ile Grande. Landed Predannack 2020.

Blue II (Sgt. Deuntzer) was on left of the formation. Attacked gun-posts from E side firing a 4-secs. burst. Seeing no results, fired a short burst at another gun-post then turned away to NE and, leaving aerodrome, fired 2 bursts at another gun-post ½ mile from the aerodrome. This ceased firing. Blue 2 saw the fire above as he left the target and at same time saw Blue 1 turn away to NE and apparently turn back towards the aerodrome. Blue 2 crossed the French coast between Les Triages and Les Sept Iles and landed Predannack 2020 hours.

Blue 1 (F/O Mackenzie) is missing. None of the 3 a/c saw any o/a in the air or on the ground nor did they see the Whirlwinds, nor results from the Whirlwinds attacks. Red 1 reports enthusiastic greetings from fishing craft in the Bay W of Lannion. The Hurricanes report the weather 5/10ths herringbone cloud at 3,000 feet. Vis. 15-20 mile 9 Spitfires landed Portreath 2010 hours. 1

Whirlwind landed Portreath 2005 hours. 1 Whirlwind landed Predannack 2005 hours. 1 Whirlwind force-landed just E of Predannack aerodrome 2030 hours. 3 Hurricanes landed Predannack 2015-2020 hours. Sgt. Hunter and 1 Whirlwind missing. F/O Mackenzie and 1 Hurricane missing. P/O Hordern slightly injured and 1 Hurricane Cat. II. 1 Whirlwind Cat. III (force-landed) near Predannack. Enemy casualties. 1 Ju88 claimed destroyed on the ground.

My thanks are due to the Public Record Office, Kew, for access to RAF Records and Reports for 501 and 247 Squadrons and RAF Kenley 1940-1941.

Index